30th Conference of the International Nuclear Target Development Society (INTDS2022)

EPJ Web of Conferences Volume 285 (2023)

Villigen, Switzerland
25 - 30 September 2022

Editors:

Dorothea Schumann
Christelle Stodel
Matthew Gott

ISBN: 978-1-7138-8042-4

Printed from e-media with permission by:

Curran Associates, Inc.
57 Morehouse Lane
Red Hook, NY 12571

Some format issues inherent in the e-media version may also appear in this print version.

For additional information, please contact EDP Sciences – Web of Conferences
at the address below.

EDP Sciences – Web of Conferences
17, Avenue du Hoggar
Parc d'Activité de Courtabœuf
BP 112
F-91944 Les Ulis Cedex A
France

Phone: +33 (0) 1 69 18 75 75
Fax: +33 (0) 1 69 28 84 91

contact-edps@webofconferences.org

Additional copies of this publication are available from:

Curran Associates, Inc.
57 Morehouse Lane
Red Hook, NY 12571 USA
Phone: 845-758-0400
Fax: 845-758-2633
Email: curran@proceedings.com
Web: www.proceedings.com

TABLE OF CONTENTS

The 30th Conference of the International Nuclear Target Development Society, INTDS 2022

Dorothea Schumann[1]*, *Matt* Gott[2], and *Christelle* Stodel[3]

[1]Paul Scherrer Institut Villigen, Forschungsstrasse 111, 5313 Villigen, Switzerland
[2]Oak Ridge National Laboratory, 1 Bethel Valley Road, Oak Ridge, Tennessee, 37830, USA
[3]Grand Accélérateur National d'Ions Lourds (GANIL) Bvd Henri Becquerel - F-14076 CAEN - France

The International Nuclear Target Development Society (INTDS) is dedicated to promoting work and networking on the fabrication and characterization of various kinds of targets for experiments in nuclear physics and related fields. Its 30th conference (INTDS2022) was held from September 25th to September 30th, 2022, at the Paul Scherrer Institute Villigen (PSI), Switzerland.

After two years of restrictions due to the pandemic, we were happy to be able to host the conference in person. In a few exceptional cases, e.g. for colleagues who were still not allowed to travel, the hybrid mode was offered.

The scientific program included contributions on the development of new target preparation methods, advances in target characterization, recent results in the preparation of targets for medical applications, gas targets, beam stripper films, targets for ion beams and much more (https://indico.psi.ch/event/7834/). It is particularly gratifying that again this year we managed to set up a session on new developments in high performance target technology.

In addition to the broad field of stable targets, covered in two sessions, radioactive targets play an increasing role in many areas of nuclear research. In addition to these radioactive targets, there is a rapidly growing demand for the availability of radioactive source material. These specific needs were addressed in several sections.

INTDS places particular emphasis on the promotion of young scientists. This concern was reflected both in financial support and in the large number of contributions from students. To underline the importance of teaching, a special session dedicated on educational topics was established.

78 participants from Africa, Asia, Europe and North America attended the conference. 51 oral contributions were presented in 12 sessions. There were four invited presentations by experts from the United States, Canada and Romania.

Numerous colleagues helped to make the conference a success. In particular, we acknowledge the work of the PSI facility managing service, the printing office, the photographer, the visitor centre, the canteen team as well as the colleagues from PSI West who performed a guided tour through the accelerator facilities.

Special thanks go to the members of the scientific advisory board for carefully selecting the contributions and to all the referees who helped improving the manuscripts for these proceedings.

Photo of the participants of the 30th INTDS conference

* Corresponding author: dorothea.schumann@psi.ch

Scientific Advisory committee

Klaus Eberhardt (JGU, Mainz, Germany)
Christelle Stodel (GANIL, Caen, France)
John Greene (Argonne National Laboratory, USA)
Ntombizonke Kheswa (iThemba LABS, South Africa)
Hironori Kuboki (RIKEN, Saitama, Japan)
Bettina Lommel (GSI, Darmstadt, Germany)
Goedele Sibbens (EC-JRC, Geel, Belgium)
Dannie Steski (BNL, Brookhaven, USA)
Yoshio Yamazaki (J-PARC Center, Tokai, Japan)
Mikael Jensen (DTU, Copenhagen, Denmark)
Daniela Kiselev (PSI, Villigen, Switzerland)

Local Organisation Committee

Dorothea Schumann (Chair)
Emilio Andrea Maugeri
Zeynep Talip
Rugard Dressler
Kiselev Daniela
Angela Blattmann
Mario Aaron Veicht
Noemi Cerboni

Participants

Maruf Abubakar	Ghana Atomic Energy Commission
Ernst Artes	JGU Mainz
Powell Barber	Florida State University
Theo Bigourdan	GIP ARRONAX
Saverio Braccini	University of Bern, Switzerland
Philipp Brand	WWU - Münster
Guillaume Brunet	GANIL CNRS
Sizwe Buwa	iTemba labs
Arnaud Cadiou	SUBATECH IN2P3
Noemi Cerboni	PSI - Paul Scherrer Institut
Nadine Chiera	PSI - Paul Scherrer Institut
Jennifer Conner	Oak Ridge National Laboratory
Stefania Dede	Texas A&M University, University of Notre Dame
Christoph Düllmann	JGU Mainz / GSI Darmstadt / HIM Mainz
Klaus Eberhardt	JGU Mainz
Hanna Eick	WWU Münster
Raffaele Esposito	CERN
Cherlyn Foster	Oak Ridge National Laboratory
Matt Gott	Oak Ridge National Laboratory
Alexander Gottberg	TRIUMF
Steffen Happel	TrisKem International
Stephan Heinitz	SCK CEN
Jukka Jaatinen	University of Jyväskylä
Vinder Jaggi	Micromatter Technologies Inc.
Pascal Jardin	CNRS/IN2P3/GANIL
Takuji Kanemura	FRIB, MSU East Lansing
Jolanta Karpinska	JRC EC
Ntombizonke Kheswa	iThemba LABS/NRF
Birgit Kindler	GSI Darmstadt
Daniela Kiselev	PSI - Paul Scherrer Institut
Bettina Lommel	GSI Darmstadt
Ashabari Majumdar	University of Notre Dame
Christian Mannweiler	WWU Münster
Athena Marenco	Los Alamos National Laboratory
Antonio Massara	INFN-LNS Catania
Michel Matthieu	GANIL
Connor Mohs	Argonne National Laboratory
Florea Nicoleta-Mihaela	IFIN-HH
Anand Pandey	University of Delhi
Franck Pérocheau	GANIL
Federico Pinna	POLITECNICO DI TORINO - DISAT
Julien Piot	GANIL / CNRS
Russell Polson	Goodfellow
Lauren Reed	JGU Mainz
Dennis Renisch	JGU Mainz
Damoy Samuel	GANIL
Matus Sedlak	LNL INFN
Gregory Severin	MSU, East Lansing
Goedele Sibbens	JRC-Geel
Thomas Sounalet	CNRS Subatech
Ingo Spahn	Forschungsgzentrum Jülich
Christelle Stodel	GANIL
Anna Stolarz	University of Warsaw
Constance Stoner	The Arizona Carbon Foil Co. Inc.
Ngoc Duy Trinh	Orano
Sean Vanbergen	TRIUMF
Nick van der Meulen	PSI - Paul Scherrer Institut
Sophia Vestrick	WWU Münster
Wim Weterings	CERN
Mike Zach	Oak Ridge National Laboratory
Robert Eichler	PSI - Paul Scherrer Institut
Dorothea Schumann	PSI - Paul Scherrer Institut
Rugard Dressler	PSI - Paul Scherrer Institut
Mario Veicht	PSI - Paul Scherrer Institut
Steffen Taut	TU Dresden
Zeynep Talip	PSI - Paul Scherrer Institut
Kristian Myhre	Oak Ridge National Laboratory
Emilio Maugeri	PSI - Paul Scherrer Institut
Jasmin Huber	PSI - Paul Scherrer Institut
Silvan Christ	Linktronix
Jason Jackson	Goodfellow
Jeri Winkleblack	accompanying person
online:	
Hiro Hasebe	Nishina Center for Accelerator-Based Science, RIKEN
Katharina Domnanich	FRIB, MSU, East Lansing
Tom Leadbeater	iTemba labs
Maria Fisichella	INFN - Laboratori Nazionali del Sud

The conference was sponsored by

edp sciences

EPJ Web of Conferences

Statement of Peer review

In submitting conference proceedings to EPJ Web of Conferences, I certify to the Publisher that I adhere to the **Policy on Publishing Integrity** of the journal in order to safeguard good scientific practice in publishing.

1. All articles have been subjected to peer review administered by the proceedings editors.
2. Reviews have been conducted by expert referees, who have been requested to provide unbiased and constructive comments aimed, whenever possible, at improving the work.
3. Proceedings editors have taken all reasonable steps to ensure the quality of the materials they publish and their decision to accept or reject a paper for publication has been based only on the merits of the work and the relevance to the journal.

Title, date and place of the conference

30th Conference of the International Nuclear Target Development Society
INTDS 2022
25 – 30 September 2022
Paul Scherrer Institut Villigen PSI, Switzerland

Proceedings editor(s):

Dorothea Schumann
Christelle Stodel
Matthew Gott

Date and editor's signature

Villigen, 16.10.2022

Thickness and uniformity analysis of thin and heat-resistant targets

Maria Fisichella[1*], *Franck* Delaunay[1,2,3], *Vittoria* Capirossi[4] , *Federico* Pinna[4], *Daniela* Calvo[5], *Matteo* Campostrini[6], *Francesco* Cappuzzello[1,2], *Manuela* Cavallaro[1], *Felice* Iazzi[5], *Antonio* Massara[1], *Valentino* Rigato[6], and *Martina* Ursino[1]

for the NUMEN collaboration

[1]Istituto Nazionale di Fisica Nucleare, Laboratori Nazionali del Sud, Catania, Italy
[2]Dipartimento di Fisica e Astronomia "Ettore Majorana", Università di Catania, Catania, Italy
[3]LPC Caen, Normandie Université, ENSICAEN, UNICAEN, CNRS/IN2P3, Caen, France
[4]Department of Applied Science and Technology, Politecnico di Torino, Torino, Italy
[5]Istituto Nazionale di Fisica Nucleare, Sezione di Torino, Torino, Italy
[6]Istituto Nazionale di Fisica Nucleare, Laboratori Nazionali di Legnaro, Padova, Italy

Abstract. The thickness and uniformity characterisation of the first prototypes of thin tellurium and germanium targets evaporated on Highly Oriented Pyrolytic Graphite for NUMEN project is reported. The contribution of such targets to the energy resolution on reaction ejectiles is evaluated.

1 Introduction

The NUMEN (NUclear Matrix Elements for Neutrinoless double β decay) project proposes an innovative technique to access experimental information on Nuclear Matrix Elements (NMEs) involved in neutrinoless double β decay (0νββ), by measuring the differential cross section of heavy-ion induced Double Charge Exchange (DCE) reactions. The idea to use DCE reactions to obtain information on NME of 0νββ is based on the fact that the two processes are characterized by important similarities, as for example the same initial and final nuclear states, and a similar mathematical structure of the transition operators [1,2].

So far, the NMEs have been calculated using different theoretical models, but the discrepancies between the values from the various calculations are large. In addition, experimental constraints are loose. The main goal of NUMEN is to establish an alternative experimental path for the determination of the 0νββ NMEs. An experimental campaign has already started by using targets made with 0νββ candidate isotopes [1]. The experiments are performed at INFN - Laboratori Nazionali del Sud (LNS, Catania, Italy) mainly with ^{18}O and ^{20}Ne beams produced by the LNS cyclotron at 15-60 AMeV and with the MAGNEX spectrometer to detect the reaction products. Since the DCE reactions are characterised by very small cross-sections, an improvement of the LNS cyclotron is underway to enhance the intensity of the beams up to 10^{13} pps. Consequently, an upgrade of the whole detection apparatus is also ongoing [3-5] to withstand these intense beams. In particular, a new target system, capable of dissipating the power released by these beams is under study [4].

In the following, the requirements for NUMEN targets are discussed. Then, the characterisation of tellurium and germanium target prototypes is described in detail. On the basis of thickness and non-uniformity measurements, the impact of the targets on the energy resolution of the ejectile is evaluated.

2 New target system for NUMEN

The power released by the intense beam in the NUMEN targets is of the order of 1 W, two orders of magnitude higher than the typical power released in a target for nuclear experiments. Analytical and numerical thermal calculations have been performed with the MATLAB framework, which allows to conclude that a stand-alone target is not able to dissipate this amount of power [6]. The heat dissipation can be enhanced by depositing the target material on a highly thermally conductive substrate, i.e. High Oriented Pyrolytic Graphite (HOPG), characterised by a thermal conductivity which could range between 1700 and 1950 W m^{-1} K^{-1} [7-8]. It has been estimated that a 450 μg/cm^2-thick HOPG is able to dissipate up to 10 W [9]. Moreover, a dedicated cooling system has also been designed to help the heat dissipation. The target will be encased in a copper holder, mounted on top of a cryocooler and kept at 40 K [3]. Thanks to the HOPG and associated cooling system, the target reaches, in a few hundred microseconds, a

* Corresponding author: fisichella@lns.infn.it

maximum temperature lower than the melting point [9,10]. Thus, the NUMEN targets will be characterised by the presence of a substrate thicker than a usual target backing. Thick HOPG substrates could produce significant straggling and cause a degradation of the energy spread of the products. To distinguish the ground state from the ground state of the DCE transition, the energy resolution of the reaction products should be lower than the energy of the first excited state of the residual nuclei, typically ≈ 500 keV. In order to obtain this resolution, and because of the thick HOPG backing, the target material thickness (i.e. the tellurium and germanium layer) should range between 250 and 500 μg/cm^2. The presence of non-uniformities in the HOPG and/or in the target can also contribute to worsening the energy resolution. For this reason, as will be explained in the following, a detailed analysis of the non-uniformity has to be performed. The maximum acceptable non-uniformity has to be estimated for each reaction considered.

3 Tellurium and germanium target production and analysis

First depositions of tellurium and germanium have been produced with an electron-beam evaporation source at the TRUSTECH Company laboratories (Turin, Italy), specialized in micro and nanotechnology. Natural tellurium and germanium were used to save costly isotopes. To define the best parameters for the evaporation (see [10,11] for details), several attempts have been made by varying the evaporation rate, by changing the HOPG substrate thickness and temperature, and using a deposition buffer for helping the diffusion of the evaporated material onto the backing surface. For the materials considered in the present work, the best results were obtained by performing the evaporation process at standard conditions, i.e. without heating of the backing, without a deposition buffer and by applying a low evaporation rate (0.2-0.5 Å/s). For the mentioned depositions, the HOPG substrate (nominal thickness: 450 μg/cm^2) was obtained from the OPTIGRAPH Company (Berlin, Germany). It has to be underlined that the HOPG substrate is always analysed before the evaporation using the alpha-particle transmission technique, to determine its actual average thickness and uniformity. This procedure showed that the HOPG foils used for these first studies have a significant non-uniformity, both globally and locally (i.e. inside the alpha spot). The average thickness could differ from the nominal value by up to 50%, with a local non-uniformity variation from 5% to 25%. Because of these characteristics, these kinds of HOPG foils cannot be used as backing for the final NUMEN targets. However, they have been useful for the definition of the evaporation procedure. A detailed characterisation of the HOPG foils will be reported elsewhere. The here presented target prototypes have been characterised using Field Emission Scanning Electron Microscope (FESEM), Rutherford Backscattering Spectrometry (RBS) and Alpha Particle Transmission (APT). FESEM provides information on the morphology of the sample surface. The analyses have been performed with a Zeiss Merlin instrument , hosted in the laboratories at the Department of Applied Science and Technology of the Polytechnic of Turin, with a magnification of up to one million and a resolution of 0.8 nm.

The RBS technique gives information on the thickness, and on the elemental composition of the sample. One of the advantages of the RBS technique is that the study of an evaporated film is not influenced by the type of sample substrate used. RBS measurements have been performed at the AN2000 accelerator facility of the INFN – Laboratori Nazionali di Legnaro (Padova, Italy). This facility provides proton and alpha beams with energies up to 2 MeV with a circular spot diameter of around 1 mm (Full Width at Half Maximum). The proton beam has been exploited to study the HOPG backings, whilst for studying the target depositions, the 2 MeV alpha beam was used. The scattered ions were detected with a silicon detector at an angle of 160° with respect to the beam. The uncertainty of these measurements has been estimated around 10%.

The APT method was used to characterise the sample in terms of thickness and uniformity, with the set-up hosted at the Department of Applied Science and Technology of the Polytechnic of Turin [11]. The average thickness value was deduced from measuring the energy lost by an alpha particle, emitted by an ^{241}Am source, in traversing the target. In this procedure, the main source of uncertainty is given by the stopping power values obtained with the SRIM code [12]. According to the comparisons of SRIM calculations with measured stopping powers [12], the uncertainty of the calculations, for the materials studied in the present work, is in the order of 5%. The target non-uniformity can be deduced, as described in [13], by comparing the Full Width at Half Maximum (FWHM) of the experimental spectrum with the FWHM of a simulated spectrum, obtained by assuming a uniform target with the measured average thickness and convoluting it using the detector energy resolution as input. In the present analysis, the simulations were performed with the SRIM [12] program.

3.1 Tellurium target

Figures 1-2 show the results of the FESEM, RBS and APT analyses of a tellurium target evaporated at the standard conditions previously mentioned. In the FESEM image (Figure 1a), the presence of tellurium grains, with dimensions in the order of tens of nanometers can be observed. The RBS analysis (Figure 1b) allows to determine the target thickness (322 ± 32 μg/cm^2) and shows that no contaminants are present in the target. Figure 2 shows the residual energy spectra for the alpha-particles passing through the HOPG backing alone (top) and the HOPG backing + tellurium (bottom). The black line is the experimental spectrum and the red and green lines are the simulated spectra used to determine the non-uniformity inside the alpha-particles spot (also referred to as "local non-uniformity"), as described in [13].

https://doi.org/10.1051/epjconf/202328501001

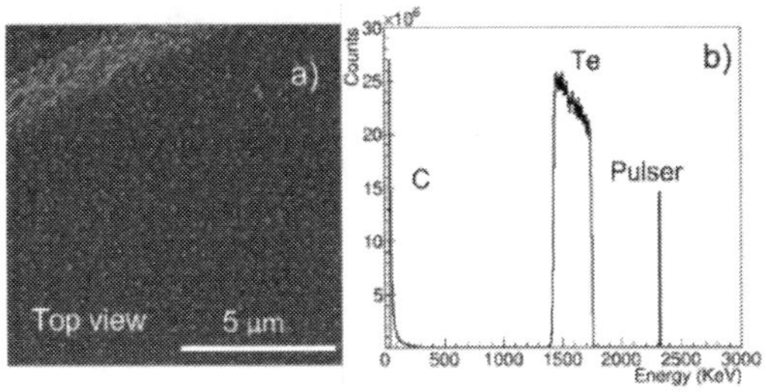

Fig. 1. (a) FESEM view of Te sample, deposited on a HOPG substrate at room temperature without any buffer (b) RBS spectrum of the Te target, measured with a 2.2 MeV alpha beam, detected at an angle of 160° with respect to the incident beam (see text for details).

According to this analysis, the HOPG substrate resulted in a thicker than the nominal value by about 20% (533 ± 27 µg/cm²) and presented a large local non-uniformity of 24%. Regarding the thickness of the deposited tellurium, the value deduced by APT (323 ± 16 µg/cm²) is in excellent agreement with the one obtained with RBS. The simulated spectrum (green curve) already includes the HOPG backing non-uniformity, thus the difference in the FWHM of the simulated and the experimental spectra is attributed to the tellurium non-uniformity, which is estimated to be about 10%.

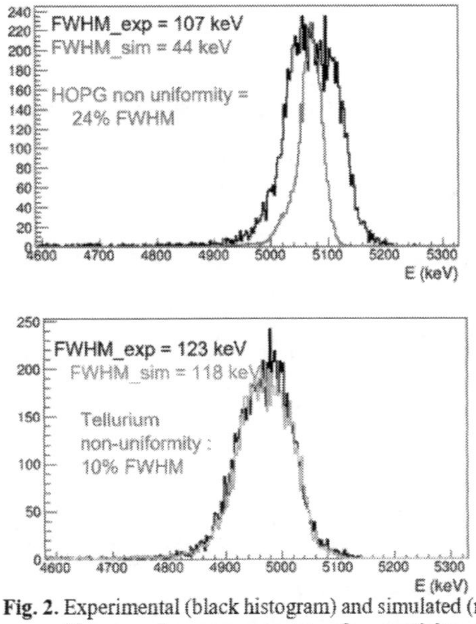

Fig. 2. Experimental (black histogram) and simulated (red and green histograms) energy spectra of α particles passing through a HOPG foil (top) and through a HOPG + Te target (bottom).

3.2 Germanium target

In the present section, similar analyses as discussed for the tellurium target are reported for a germanium target prototype produced at standard conditions. The FESEM image (Figure 3a) shows a very flat and uniform germanium deposition, with the presence of some particles with a diameter of about 0.5 micrometers.

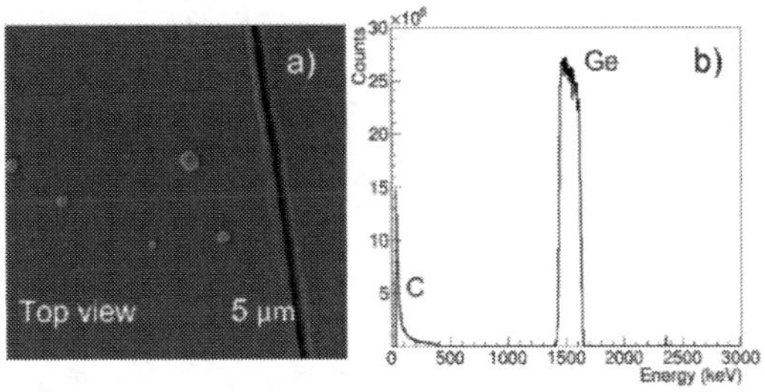

Fig. 3. (a) FESEM view of Ge sample, deposited on a HOPG substrate at room temperature without any buffer (b) RBS spectrum of the Ge target, measured with a 2.2 MeV alpha beam, detected at an angle of 160° with respect to the incident beam (see text for details).

Also in this case, the RBS analysis (Figure 3b) confirms the high purity of the deposited material, and the germanium thickness deduced from this analysis (142 ± 14 µg/cm²) is in good agreement with the one obtained from APT measurements (152 ± 8 µg/cm²). According to the alpha spectra analysis (Figure 4), the HOPG thickness is 496 ± 25 µg/cm² (i.e., 10% thicker than the nominal value) and the local non-uniformity was estimated to be about 10%. Instead, the germanium non-uniformity resulted less than 10%, in agreement with the fact that the surface seems flatter than the tellurium one.

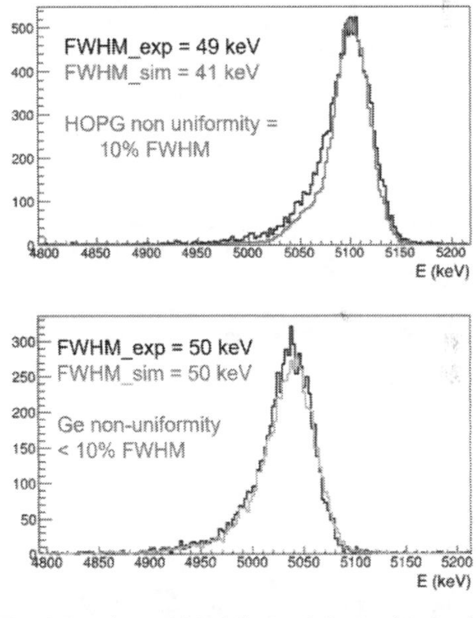

Fig. 4. Experimental (black line) and simulated (red and green lines) energy spectra of α particles passing through a HOPG foil (top) and through a HOPG + Ge foil (bottom).

4 Energy resolution estimations

To judge the quality of the target prototypes, analytical calculations were performed to estimate the energy resolution of an ejectile, detected at 0° with respect to the incident beam, produced in a DCE reaction in the considered target. For the tellurium case, the reaction ^{130}Te(^{20}Ne,^{20}O)^{130}Xe was considered. The first excited state of ^{130}Xe has an energy of 536 keV, and therefore, the energy resolution of the ejectile needs to be below this value. The calculations take into account the contribution of the cyclotron and the MAGNEX spectrometer, as well as the target effects (i.e. dispersion due to the random reaction point in the Te film, energy straggling in the Te film and HOPG backing, non-uniformity of the Te layer and HOPG substrate). Since calculations do not take into account other factors which contribute to energy spread (e.g. statistical counting, peak shape, presence of background, etc.), it has been estimated, thus to have a safety margin, that the energy resolution value should be at least 10% less than the energy gap. For a target with the characteristics of the Te prototype (^{130}Te film 323 μg/cm^2 thick with a non-uniformity of 10% on a HOPG backing 533 μg/cm^2 thick with a non-uniformity of 24%), the ejectile total energy resolution has been estimated to be 525 keV, which is very close to the energy of the first excited state. To investigate the role played by the backing non-uniformity, a series of calculations was performed by assuming a 323-μg/cm^2 ^{130}Te thick deposited on a 450 μg/cm^2 HOPG substrate with a non-uniformity ranging from 5% to 25%. The energy resolution values calculated as a function of the HOPG non-uniformity are given in Figure 5 (orange points). The plot also shows the lower limit of the energy resolution due to the cyclotron and the spectrometer, as well as the energy of the ^{130}Xe first excited state. A second set of calculations has also been performed, assuming a Te non-uniformity of 20% (blue squares). It is clear that the non-uniformity of the HOPG substrate contributes more to the energy resolution than the tellurium layer.

The same type of analysis was performed for the Ge target described in the previous section. In this case, the considered reaction was ^{76}Ge (^{20}Ne,^{20}O)^{76}Se. The ^{76}Se first excited state has an energy of 559 keV. For a ^{76}Ge target with the characteristics of the prototype (Ge thickness 153 μg/cm^2 with an uniformity of 10% on a 496 μg/cm^2 thick HOPG backing with a non-uniformity of 10%), the ejectile energy resolution has been estimated to be 445 keV, which is below the maximum acceptable value for the energy resolution. Given these target characteristics, it has been estimated that increasing the Ge thickness to 300 μg/cm^2 would still give an acceptable energy resolution (477 keV).

5 Conclusion and outlook

Here, the characterisation of the first tellurium and germanium target prototypes for the NUMEN project was described. Calculations to estimate the total energy resolution of DCE reaction products expected from such targets were performed. Results show that the non-uniformity of the HOPG backing plays a very important role. According to the results presented in the previous section, for Ge and Te targets, a 450-μg/cm^2 HOPG with maximum value of non-uniformity of 20% can be used. However, a HOPG backing with a non-uniformity smaller than 10% is preferred, in order to maintain a resolution close to the limit reachable with a uniform target+HOPG assembly. This would help disentangling

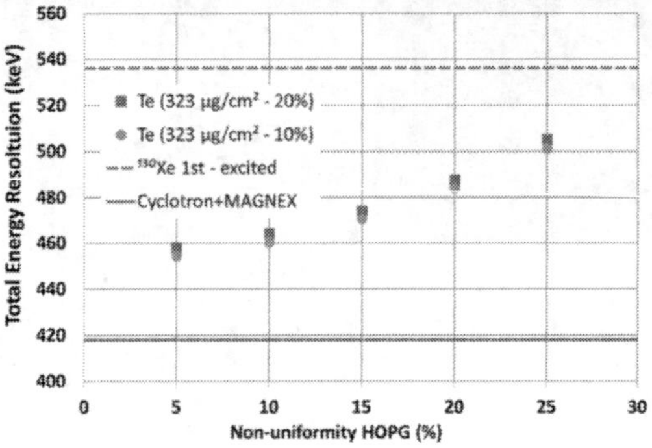

Fig. 5. Total energy resolution versus HOPG non-uniformity, for the ejectile of the reaction ^{130}Te(^{20}Ne,^{20}O)^{130}Xe, detected at 0° (see details in text). The thickness of the HOPG is 450 μg/cm^2.

energy peaks also in the presence of factors not taken into account in the calculations. For this reason, new HOPG substrates provided by different companies are under study. Concerning the depositions of tellurium and germanium on HOPG supports satisfactory uniformities have been obtained for these materials at standard evaporation conditions. A drawback of the evaporator used in the present study was the large amount of material required (from 10 to 20 g) to perform the evaporation, due to geometrical reasons. Given that the final NUMEN targets will be made of isotopically enriched material, it is necessary to reduce this quantity by at least an order of magnitude. For this reason, a new series of evaporations is on-going at LNS, with the aim of producing target prototypes with characteristics similar to the ones described in this work, using though amounts of material lower than 1 g.

References

1. F. Cappuzzello et al., Eur. Phys. J. A (2018) **54**: 72
2. F. Cappuzzello et al., Progress in Particle and Nuclear Physics 128 (2023) 103999
3. F. Cappuzzello et al., International Journal of Modern Physics A 36, 30 (2021) 2130018
4. D. Calvo, J. Phys. Conf. Ser, 2374, 012029 (2022)
5. D. Calvo, Nucl.Instr.Meth. A 1041, 167336 (2022)
6. F. Iazzi et al, WIT Trans. Eng. Sci. 116, 61 (2017)
7. http://www.optigraph.eu/basics.html
8. https://industrial.panasonic.com/cdbs/www-data/pdf/AYA0000/AYA0000C59.pdf
9. Pinna et al., Acta Polonica B, 51, 3 (2020)
10. F. Pinna, PhD Thesis, Politecnico di Torino (2019) [https://iris.polito.it/handle/11583/2729322]
11. V. Capirossi, PhD Thesis, Politecnico di Torino (2021) [https://iris.polito.it/handle/11583/2897010]
12. J. Ziegler, "SRIM code" [http://www.srim.org]
13. M. Fisichella et al, Phys. Rev. C **92** 064611 (2015)

DICE: Apparatus for Detection of Internal Conversion Electrons

Lauren E. Reed[1*], *Dennis* Renisch[1,2], and *Christoph E.* Düllmann[1,2,3]

[1]Department Chemie - Standort TRIGA, Johannes Gutenberg-Universität Mainz, 55099 Mainz, Germany
[2]Helmholtz-Institut Mainz, 55099 Mainz, Germany
[3]GSI Helmholtzzentrum für Schwerionenforschung GmbH, 64291 Darmstadt, Germany

Abstract. An apparatus for the Detection of Internal Conversion Electrons (DICE) has been built and commissioned at Johannes Gutenberg University Mainz (JGU) for the study of recoil ion sources of ^{239}Pu, which emit ^{235m}U nuclei that deexcite to the ground state by emitting low-energy internal conversion (IC) electrons. We present an overview of DICE and its commissioning with ^{207}Bi and ^{235m}U electron sources, demonstrating that DICE allows the detection and counting of IC electrons from ^{235m}U deexcitation. Our preliminary ^{235m}U half-life of ≈ 26 min agrees with literature. DICE is thus an interesting tool to broaden analytical capabilities for recoil ion source characterization via studies of the peculiar decay of ^{235m}U.

1 Introduction

Ultra-low-lying isomers have excitation energies of < 100 eV, rather than more typical isomers with energies of tens to hundreds of keV. These ultra-low-lying isomers, which bridge nuclear and atomic physics and are of interest in the particle physics community, have recently become a topic of great interest, mainly due to the lowest known isomer, ^{229m}Th and its potential to develop nuclear clocks that possibly outperform current atomic clocks by orders of magnitude by employing nuclear transitions rather than transitions in the electron shell [1-2]. One such ultra-low-lying isomer is ^{235m}U, the second-lowest known isomeric state, located at 76.8(5) eV excitation energy [2-4]. ^{235m}U ($I^{\pi} = 1/2^{+}$) is the first excited state of ^{235}U ($I^{\pi} = 7/2^{-}$) and deexcites by a highly converted (conversion coefficient $\alpha \approx 10^{21}$) E3 transition to ^{235}U [5]. The internal conversion (IC) process usually involves inner subshell electrons, but if the nuclear transition energy is very small, the inner subshell electrons have binding energies that exceed the isomer excitation energy. Therefore, IC must proceed in the outer electron subshells. The half-life of ^{235m}U has been reported to range from 24.7(3) min to 27.0(2) min, depending on the chemical environment, and is usually given as ≈ 26 min [6-12].

The low excitation energy of ^{235m}U implies the conversion electrons (CEs) to be emitted with very low kinetic energy. They can only travel a few atomic layers in solids [13]. One way to obtain ^{235m}U is as a recoil daughter from ^{239}Pu thin layers. These emit ^{235m}U, with the isomer being populated in the ^{239}Pu α-decay with a summed branching of 99.8% [14]. The quantification of ^{235m}U emitted from the sample allows for obtaining information on the recoil ion rate of the ^{239}Pu thin layers. The rate at which ^{235m}U becomes available for study is a key performance indicator of the recoil ion source. To determine source recoil efficiency, it is necessary to quantify the amount of $^{235(m)}$U recoil ions emitted from the source per time quanta and compare this with the theoretical amount of $^{235(m)}$U recoil expected in the same time quanta. While the ^{235}U ground state has a too large half-life ($7 \cdot 10^{8}$ a) to be observed via α-spectrometry, measurements of the isomer decay proceeding with ≈ 26 min half-life are highly sensitive. To efficiently detect the corresponding low-energy CEs, the Detection of Internal Conversion Electrons (DICE) setup has been built and commissioned at JGU. For the present work, a 22 mm diameter ^{239}Pu thin layer recoil ion source was prepared by the molecular plating (MP) technique [15]. Recoiling $^{235(m)}$U nuclei can be collected in a suitable catcher, from which CEs can emerge. From the rate of emitted CEs, the number of implanted ^{235m}U nuclei can be determined, thus allowing the calculation of the ^{239}Pu thin layer recoil efficiency. The present work provides a technical overview of DICE and the first commissioning results.

2 DICE – Detection of Internal Conversion Elections

2.1 General setup

The DICE setup (Fig. 1) comprises a vacuum chamber evacuated by a turbomolecular pump (Agilent Turbo-V 511 Navigator) to ≈ $1 \cdot 10^{-6}$ mbar within 5 min. A Pirani gauge monitors the pressure. The chamber houses a two-stage multi-channel plate (MCP) stack and an insulated ^{235m}U source holder placed directly in front of the stack. The source holder is connected to a rod, which exits the chamber and allows manipulation of the source-to-MCP distance by up to 25 mm. The holder accommodates samples of up to 22 mm diameter. The source can be

* Corresponding author: lreed@uni-mainz.de

 https://doi.org/10.1051/epjconf/202328501002

biased up to -2 kV using a high voltage (HV) power supply (NHQ 203M) to propel the low-energy CEs. Impulses detected from the MCP stack are decoupled, amplified and discriminated. The signal is then sent to a field-programmable gate array (FPGA) board, which computes the data and displays them in the chosen format on the computer.

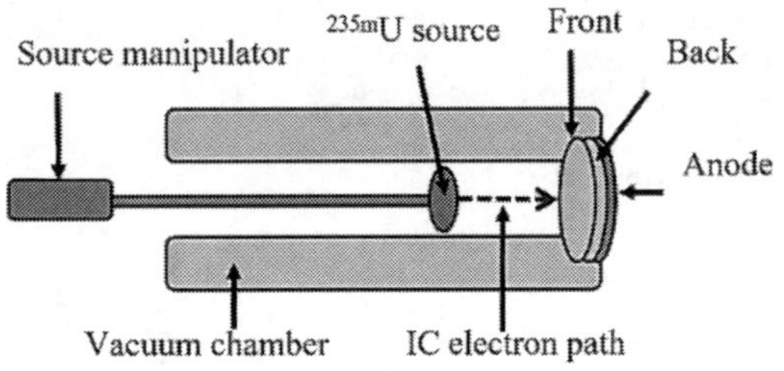

Fig. 1. Schematic of the Detection of Internal Conversion Electrons (DICE) apparatus.

2.2 Multi-channel plate (MCP) stack

The MCP stack (DET40 MCP detector with timing anode, RoentDek Handels GmbH) is mounted inside the vacuum chamber flange. It has a chevron-style configuration (two-stage stack) with a 12 µm pore diameter and can be used to count individual particles such as electrons, photons and ions with respect to time. Each plate features a 45 mm active diameter, a thickness of 1 mm, a bias angle of 20°, and an open area ratio of 70 %. The operating pressure is $< 2\cdot 10^{-6}$ mbar, and the maximum count rate is ~1 MHz. The front plate was always grounded, the back plate (B) was biased to +1900 V, and the anode (A) to +2100 V to ensure attraction and acceleration of the electrons. Later tests had B biased to +2200 V and A to +2500 V.

2.3 Control and acquisition systems

MCP signals coming from the anode were decoupled (RoentDek HFSD-BNC), pre-amplified (ORTEC VT120C +12 V fast pre-amp) and discriminated (ORTEC 584 constant fraction discriminator). Signals were counted by the FPGA (TUL PYNQ-Z2) board, which was python programmable in Jupyter notebook.

3 Commissioning of DICE

3.1 Reagents and materials

A ^{239}Pu thin layer was deposited on a 525 µm Si wafer substrate by MP [15]. A Pt coiled wire (Goodfellows, 99.9 %, 1 mm thickness) served as the anode [16]. All reagents were of high-purity chemical grade (Sigma-Aldrich). A MP cell, made from Kel-F to minimize Pu absorption, was set up with a 22 mm diameter open deposition area at the bottom of the cell [17]. Ethylene-propylene-diene-monomer (EPDM) gaskets, resistant against N,N-dimethylformamide (DMF), were used. All MP apparatus and substrates were washed with isopropanol, then acetone, and subsequently again with isopropanol before plating. After washing, the anode was additionally etched with 6 M HCl, washed with water and then isopropanol. Blank platings were performed three times before starting MP.

^{239}Pu stock solution in dilute HCl contained about 0.2 % ^{241}Am. It was prepared for ^{239}Pu plating by evaporation and redissolution with four aliquots of conc. HNO_3 and two aliquots of 0.1 M HNO_3.

3.2 ^{239}Pu source preparation and analytics

135 kBq of ^{239}Pu, dissolved in 10 µL 0.1 M HNO_3 was pipetted into 10 mL DMF, mixed, and placed in the chimney of the cell. MP was carried out for 75 min at room temperature and a constant current density of 0.9 mA/cm^2. Voltages did not exceed 400 V. One minute before the end of plating, 1 mL of conc. NH_4OH was added to the solution to prevent redissolution. The finished source was allowed to dry in the air [18]. It was characterized by radiographic imaging (FUJIFILM FLA 7000), scanning electron microscopy (SEM) at 20 kV (Philips XL30), and α-spectrometry (Canberra alpha analyst with Si p-i-n detector). After MP, the ^{239}Pu content of an aliquot of the supernatant solution was measured and compared to that of an aliquot taken before MP had begun to give the indirectly determined yield. The direct yield was obtained from an α-measurement of the source.

3.3 Collection of recoiling ^{235m}U

A 16 µm Al foil was placed 1 mm above the ^{239}Pu recoil ion source under a vacuum of $\approx 1\cdot 10^{-2}$ mbar until saturation (> 4 h). The ^{239}Pu layer was sufficiently thin to allow recoiling ^{235m}U (88 keV) to escape the layer and implant in the Al foil to a depth of ≈ 34 nm [19]. Some ^{239}Pu sputtering may also have occurred. Quick-release valves on the vacuum chamber allowed for fast (within 10 min) removal of the foil, which was then transferred to DICE for measurement, thus minimizing ^{235m}U decay losses. The geometrical collection efficiency of ^{235m}U recoils collected in the foil at saturation was calculated to be 43 % using the non-point source equation for geometrical efficiency.

3.4 DICE commissioning with ^{207}Bi

For first commissioning, a 29.5 kBq ^{207}Bi CE source ($t_{1/2}$ = 31.55 a) was mounted inside DICE. Four tests (Table 1) were performed at MCP stack bias voltages of; B: +1900 V, A: +2200 V (lower MCP bias) and B: +2200 V, A: +2500 V (higher MCP bias). Registered counts were summed over a defined interval, recorded, and then the counts reset and the process repeated for the total run time. Test no. 1 identified the optimum discriminator threshold at grounded source bias, and test no. 2 identified the optimum source bias at the optimum discriminator threshold found in test no. 1. Test no. 3 identified the optimum discriminator threshold at the optimum source bias found in test no. 2. Test no. 4 was longer runs at the optimum source bias and optimum discriminator threshold at that bias to assess system stability and experiment reproducibility. All parameters other than the independent variable were kept constant for each test. The discriminator threshold was raised in +10 mV intervals and the source bias in -100 V intervals. The background was recorded at each MCP stack bias voltage.

Table 1. ^{207}Bi source tests 1-4 at lower MCP bias and repeated at higher MCP bias. Optimum discrim. thresh. and source bias was found to be + 60 mV and – 900 V, respectively, for lower and higher MCP biases.

Test no.	Interval (s)	Time per run (s)	No. of runs	Discrim. thresh. (+ mV)	Source bias (- V)
1	10	60	13	0 – 140	0
2	10	60	15	60	0 -1400
3	10	60	13	0 – 140	900
4	10	300	2	60	900

3.5 Measurement of the ^{235m}U half-life

The ^{207}Bi results verified the correct operation of DICE. However, they cannot be directly applied to ^{235m}U as the CE energies between the two nuclides differ drastically; biasing affects ^{207}Bi CEs in the keV range much less than the ^{235m}U CEs in the eV range. Once DICE was proven to detect electrons with good repeatability and low noise, initial ^{235m}U CE measurements began. A foil with freshly collected ^{235m}U was mounted inside DICE for each run. At the lower MCP stack bias, four runs (1-4) were performed (Table 2, test no. 1). Afterwards, an α-spectrometric measurement of the foil was performed to quantify Pu sputtering during collection. At the higher MCP stack bias, six runs (A-F) were performed (Table 2, test no. 2). The discriminator threshold was varied (+ 0, 10, 20, 30, 60 and 90 mV) for testing purposes. All other parameters were unchanged.

Table 2. ^{235m}U tests at (1, runs 1-4) lower MCP bias, and (2, runs A-F) higher MCP bias.

Test no.	Interval (s)	Total time (s)	No. of runs	Discrim. thresh. (+ mV)	Source bias (- V)
1	30	>9000	4	60	900
2	30	360	6	0, 10, 20, 30, 60, 90	900

All results were corrected for decay loss between the end of collection and the start of counting. Uncertainty in counts was given as the square root of counts. Data were analyzed using a two-component exponential decay fit using a python-based least squares fitting model [20].

An average constant background determined independently was subtracted from the data sets. The point of intersection from the two best fits for each component was used to calculate the ^{235m}U half-life via $N_t = N_0 e^{-(\ln(2)/t_{1/2})t}$ where N_t is the number of ^{235}U atoms at time t, N_0 is the number of ^{235m}U atoms at $t = 0$, and $t_{1/2}$ is the half-life of ^{235m}U.

4 Results and discussion

4.1 Source preparation and characterization

Radiographic imaging of the ^{239}Pu source showed a homogeneous activity distribution. In SEM images, the source surface looked smooth and crack-free. An activity of 135 kBq, corresponding to an areal density of ≈ 15.5 μg/cm^2, and an MP yield of 96(2) %, confirming efficient MP of the ^{239}Pu recoil ion source, was deduced from α-spectrometry.

4.2 DICE commissioning with ^{207}Bi

For the ^{207}Bi source tests (Table 1), background subtracted average counts were 39(2) cps and 2455(4) cps for the lower MCP stack bias and the higher MCP stack biasing, respectively. Errors were small, indicating excellent reproducibility. Background rates, due to, e.g., electronics noise, were low, i.e., in the 3-5 cps range. The background was slightly higher at the higher MCP stack bias than at the lower one. This may be due to the increased bias voltage attracting particles inside the chamber more efficiently. This also occurred for ^{207}Bi source counts, rendering background increase insignificant. The number of electrons detected from the ^{207}Bi source increased more strongly than the background rate for the higher MCP bias, suggesting that B: +2200 V, A: +2500 V biasing is more efficient for the detection of ^{207}Bi CEs.

4.3 Measurement of the ^{235m}U half-life

Between the two MCP bias data set results (Tables 3 and 4), the preliminary average half-life of ^{235m}U deexcitation is 26.2(4) min, agreeing with the accepted value of ≈ 26 min [6-12]. Run D was excluded as an anomaly due to the lower counts recorded and its much shorter half-life (21.0(5) min). Results from runs 1-4 showed small and varying amounts of ^{239}Pu sputtering occurred during collection; this did not correlate with collection time. Fig. 2 shows the exponentially decaying fit (R^2=0.998) to the experimental data and the associated residuals for run 2. There was no correlation between discriminator threshold and half-life. The characterization of the background is ongoing and is needed to reach a data quality that is high enough to be sensitive to the variations of the ^{235m}U half-life.

Tables 3 and 4. Preliminary ^{235m}U half-life and uncertainties (1 sigma confidence level) for runs at the lower MCP bias (left) and at the higher MCP bias (right). Mean uncertainty is the weighted standard deviation.

Run no.	Half-life (min)
1	27.6 ± 0.5
2	27.8 ± 0.4
3	26.6 ± 1.0
4	28.9 ± 0.5
Mean	27.7 ± 0.9

Run no.	Half-life (min)
A	25.1 ± 0.6
B	23.8 ± 0.6
C	24.2 ± 0.5
D	21.0 ± 0.5
E	24.9 ± 0.6
F	24.9 ± 0.6
Mean	24.6 ± 0.6

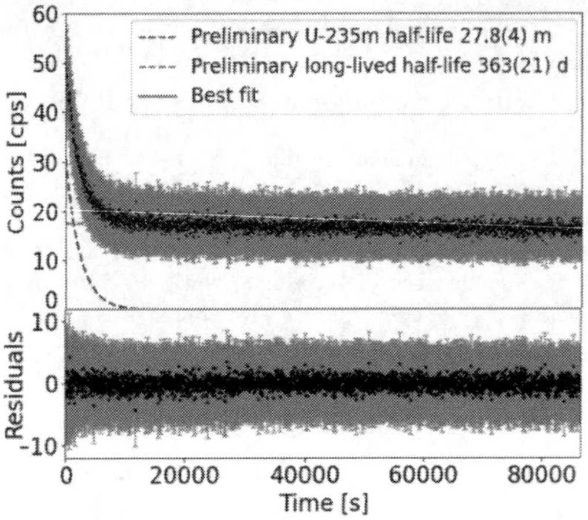

Fig. 2. Example ^{235m}U half-life determination from run 2 at the lower MCP bias. A preliminary ^{235m}U half-life of 27.8(4) m and longer-lived component of 363(21) d were obtained.

The half-life of 26.2(4) min between the two datasets agrees perfectly with 26.1(1) min obtained by Neve de Mevergnies for $^{235m}UO_3$ [7-8]. Assuming ^{235m}U in the present experiment to be mainly in the +6 state, the results of [7-8] agree with our half-life. Neve de Mevergnies obtained a ^{235m}U half-life of 25.7(3) min with V catcher foil (average Pauling electronegativity of 1.63). Although der Mevergnies did not use Al substrate (average Pauling electronegativity of 1.61), results with V foil are very close to this experiment [11]. Izawa and Yamanaka found ^{235m}U half-life to be 25.7 ± 0.4 mins, which also agrees nicely with the present study's half-life [12].

5 Conclusion and outlook

The DICE apparatus built and commissioned at JGU allows the detection and counting of CEs from ^{235m}U deexcitation. The preliminary half-life of 26.2(4) min agrees well with literature [6-12]. However, results among runs are not yet fully reproducible, with standard uncertainties in half-life of 3 % and 2 % for runs at lower and higher MCP stack bias, respectively. The higher MCP stack bias of B: +2200 V, A: +2500 V appear best for detecting both high- and low-energy electrons. DICE is thus well suited to quantify the rate of ^{235m}U emitted by ^{239}Pu-containing thin layers prepared to provide the exotic ^{235m}U for fundamental studies of its still scarcely studied properties. The short ^{235m}U half-life and correspondingly high specific activity render ^{239}Pu spiking of other α-decay recoil ion sources that emit long-lived (hence low-activity and thus hard to quantify) daughters attractive for performance studies of such sources. Upon final commissioning, DICE also promises to allow further study of the intriguing variation of ^{235m}U half-life depending on the chemical environment, which differs in literature by up to 9.5 %. Our preliminary studies suggest that after final optimization of background suppression, DICE should be sensitive to this half-life variation.

The authors thank A. Raggio and I. D. Moore for their support and continued investigations with ^{239}Pu recoil ion sources and python scripting. The authors thank S. Nagy, T. Kieck and the staff of the workshops at JGU TRIGA for their guidance in commissioning DICE. This project has received funding from the European's Union Horizon 2020 Research and Innovation Programme under grant agreement number 861198 project 'LISA' (Laser Ionization and Spectroscopy of Actinides) Marie Sklodowska-Curie Innovative Training Network (ITN).

References

1. L. von der Wense et al., Nature **533**, 47 (2016)
2. F. Ponce, E. Swanberg, J. Burke, R. Henderson, S. Friedrich, Phys. Rev. C **97**, 054310 (2018)
3. T. Sikorsky et al., Phys. Rev. Lett. **125**, 14 (2020)
4. L. von der Wense, B. Seiferle, Eur. Phys. J. A **56**, 11 (2020)
5. A.D. Panov, J. Exp. Theor. Phys. **85**, 313 (1997)
6. V.I. Zhudov, A.G. Zelenkov, V.M. Kulakov, V.I. Mostovoi, J. Exp. Theor. Phys. **30**, 516 (1979)
7. M. Neve de Mevergnies, P. Del Marmol, Phys. Lett. B **49**, 428 (1974)
8. M. Nève de Mévergnies, Phys. Rev. Lett. **23**, 422 (1969)
9. Y. Shigekawa et al., Phys. Rev. C **98**, 014306 (2018)
10. M. Neve de Mevergnies, Phys. Rev. Lett. **29**, 1188 (1972)
11. A.L. Allred. J. Inorg. Nucl. Chem. **17**, 215 (1961)
12. Y. Izawa, C. Yamanaka, Phys. Lett. B **88**, 59 (1974)
13. Y. Shigekawa, Y. Kasamatsu, A. Shinohara, Rev. Sci. Instrum. **87**, 053508 (2016)
14. E. Browne, J.K. Tuli, Nucl. Data Sheets **122**, 205 (2014)
15. W. Parker, R. Falk, Nucl. Instrum. Methods **16**, 355 (1962)
16. H. Klemenčič, L. Benedik, Appl. Radiat. Isot. **68**, 1247 (2010)
17. N. Trautmann, H. Folger, Nucl. Instrum. Methods A **282**, 102 (1989)
18. E. Bond, W.A. Moody, D.E. Dry, M.W. Rabin, J. Radioanal. Nucl. Chem. **296**, 793 (2012)
19. J.F. Ziegler, M.D. Ziegler, J.P. Biersack, Nucl. Instrum. Methods. B **268**, 1818 (2010)
20. G. Friedlander, J.M. Miller, J.W. Kennedy, *Nuclear and radiochemistry*, 2nd ed., John Wiley & Sons, New York (1961), p. 69

Cluster size determination using shadowgraphy measurements

Hanna Eick[1*], *Philipp* Brand[1], *Christian* Mannweiler[1], *Sophia* Vestrick[1], and *Alfons* Khoukaz[1]

[1]Westfälische Wilhelms-Universität Münster, Institut für Kernphysik, Wilhelm-Klemm-Straße 9, 48149 Münster, Germany

Abstract. Cluster-jet beams offer a wide variety of possible applications. The advantages of an internal and windowless cluster-jet target make it suitable for many experiments. It is important to know all the properties of the cluster-jet to prepare it for the desired conditions of the respective experiments. An essential property is the size of the clusters and their size distribution. To investigate the size of the clusters shadowgraphy measurements are performed. This article provides an overview of the shadowgraphy method and presents first results of these analyses using hydrogen clusters. These show an average cluster size of a few micrometers, which will also be of high interest, besides the here shown laser cluster interaction, for other installations using cluster beams.

1 Introduction

Cluster-jet targets developed and built by the University Münster are central components for a wide range of experiments. Some of the advantages, which make this type of target highly suitable for the future $\overline{\text{P}}$ANDA experiment at FAIR in Darmstadt, are that the target is windowless and achieves the required thickness in the range of 10^{15} atoms/cm² after a distance of more than 2 meters from the nozzle [1, 2]. To optimally adapt the cluster-jet targets to the desired conditions of every experiment, it is necessary to understand all properties of the target. One of the properties is the size of the clusters and their size distribution. This information is also important for other analyses and simulations, which require understanding of cluster-jet target properties like, e.g., the cluster evaporation process.

2 Shadowgraphy measurements at the ARCTURUS TW laser system

For the investigation of the cluster size distributions, shadowgraphy measurements have been performed at the Heinrich-Heine University in Düsseldorf. A cluster-jet target [2, 3, 4], which can be seen in fig. 1, has been mounted there to investigate the laser cluster interaction [4, 5] with the ARCTURUS TW laser system [6] in several measurement campaigns.

The target is constructed in a way, that the process gas, which is in most of the cases hydrogen, is cooled down by a two-staged cryocooler. Due to the specially shaped convergent-divergent nozzle, called Laval nozzle, clustering/atomization of the cooled gas or liquid can take place. The high mass of the clusters allows them to travel a large distance of several meters without an impact of scattering with the residual gas under usual operating conditions and therefore high densities at this distance can be reached. In contrast to the future $\overline{\text{P}}$ANDA experiment, the experiments with the ARCTURUS laser are performed only a few centimeters behind the nozzle, so additional elements for extracting the target beam and separating it from residual gas are not necessary.

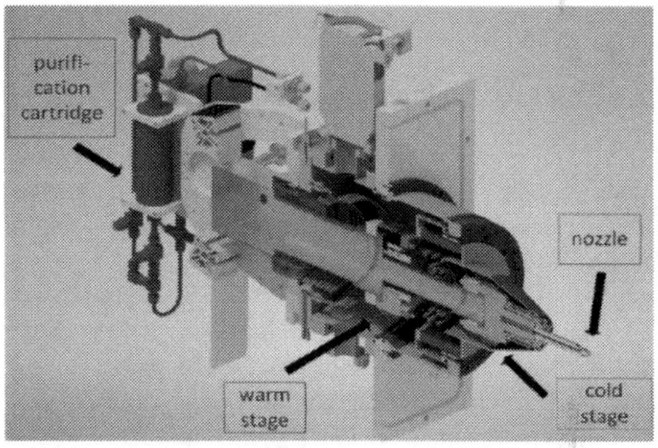

Fig. 1. CAD drawing of the cluster-jet target operated in Düsseldorf to study laser cluster interactions. The gas is purified with a purification cartridge before it enters the target, where it is cooled at the two stages of a cold head. The gas enters the nozzle at cryogenic temperatures and at high pressures and forms a cluster-jet. Figure generated by D. Bonaventura (WWU).

2.1 Setup of the measurements

The setup used for the shadowgraphy measurements can be seen in fig. 2. The image shows the experimental setup with the cluster-jet target and the ultrashort pulse laser beam. The laser beam with a wavelength of 800 nm and a pulse length of 30 fs enters the chamber, hits the cluster beam and is directed onto an objective and a subsequent camera with an exposure time much longer than the pulse length of the laser. The cluster beam is

* Corresponding author: h_eick03@uni-muenster.de

generated in the cluster source and directed to the vacuum pumps. Due to the short duration of the laser beam, the clusters can be assumed as stationary in this time window. Shadowgraphy images are taken at different target stagnation conditions with liquid hydrogen in front of the nozzle and horizontal positions in the cluster beam.

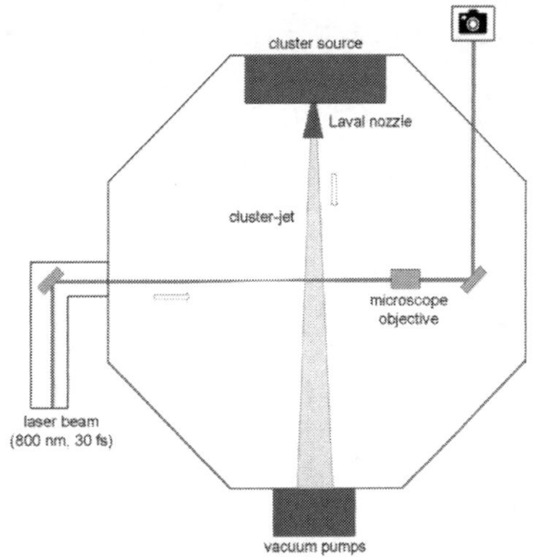

Fig. 2. Setup of the shadowgraphy measurements with the cluster-jet beam from top to bottom and the laser beam from left to right. The shadowgraphy images are taken with a CCD camera outside of the chamber.

2.2 Shadowgraphy image

With the setup shown in fig. 2, a large number of shadowgraphy images have been taken. An example for a background subtracted shadowgraphy image is shown in fig. 3. Therefore, background images are taken at a gaseous target condition, where the clusters are smaller and thus not visible. In the picture one can see three sharp clusters, that are visible as dark, slightly smeared out spots. Some disturbing structures,

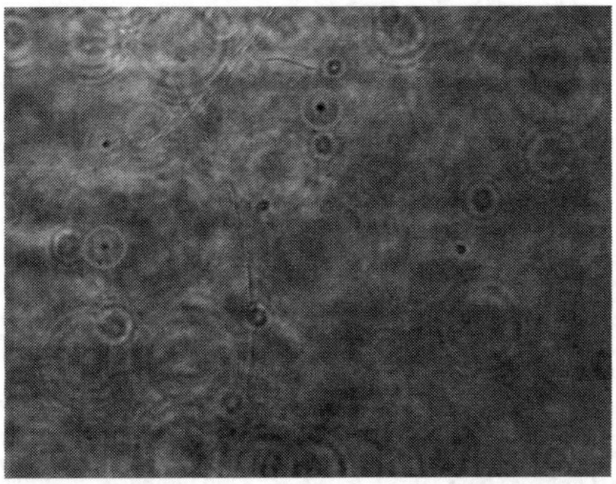

Fig. 3. Shadowgraphy image of a cluster beam with three dark, slightly smeared out clusters. Additional structures on the picture are due to other clusters outside of the focal volume or due to dust on the optical components.

like e.g., interference rings of clusters, that are too far away from the focal plane, are apparent and some dust on the camera and the optics still remains despite the background subtraction due to the shot-to-shot variation of the laser beam.

2.3 Determination of cluster size distribution

To determine the size distribution of the clusters an algorithm has been developed, which finds potential cluster candidates and decides via cluster selection criteria whether it is a cluster or not. After this sorting process all remaining cluster-like structures are fitted with a two-dimensional function and the diameter is extracted. The resulting cluster size distribution can be seen in blue in fig. 4.

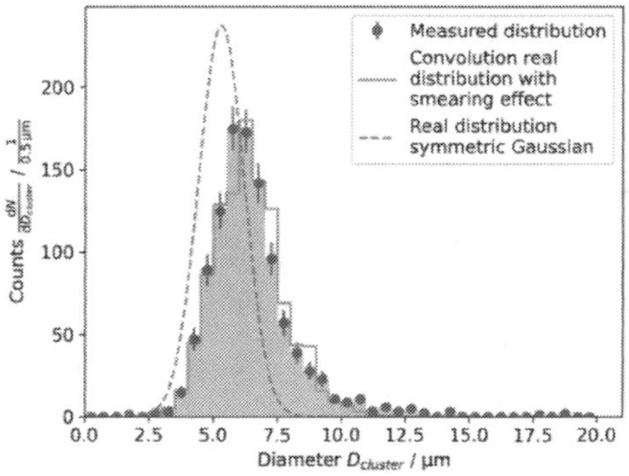

Fig. 4. Measured raw cluster size distribution for a target setting of 24 K and 16 bar and a fixed position in the cluster beam (blue). To correct for imaging effects a deconvolution is performed, which is shown in green.

In the analysis of the data, it has been found that the extracted size of the clusters depends on the individual distance between the cluster to be examined and the focal plane of the microscope objective. A deconvolution is made to account for this optical imaging effect. The real cluster size distribution can be described well by a Gaussian function, which is shown in fig. 4 in green. To approximate the raw data (blue histogram), a convolution of this Gaussian function with a smearing effect (orange histogram), caused by the mentioned out-of-focus plane observation, is fitted to the data in the blue histogram. The convolution describes the measured data sufficiently well. With the shadowgraphy method, clusters are found in the range from 3 to approx. 10 μm. The latter value establishes an upper limit on the maximum cluster size. The course of the measured cluster size distribution (blue curve in fig. 4) shows a lower boundary limited by resolution. This resolution limit can be seen as a steeper trend on the left-hand side of the distribution and is caused by the wavelength of the laser and the numerical aperture of the microscope objective. To get more insight into the distribution of possibly smaller and with this method invisible clusters, different methods have to be tested [7, 8].

2.4 Determination of volume density

Using the evaluated cluster size distributions, it is possible to determine a volume density and to compare what percentage of the gas flowing through the nozzle is contained in the visible clusters analyzed with the shadowgraphy measurements. Fig. 5 shows the measured volume density based on the shadowgraphy results in blue under the usage of the mean volume of a cluster, the number of all clusters at a measured setting, the liquid hydrogen density and the volume of the images. The expected volume density in orange is calculated from the hydrogen flow, which passes the nozzle and the cluster velocity that can be measured via a ToF setup [9].

As the data analysis is not yet completed, no final quantitative conclusion can be drawn from fig. 5 at this stage.

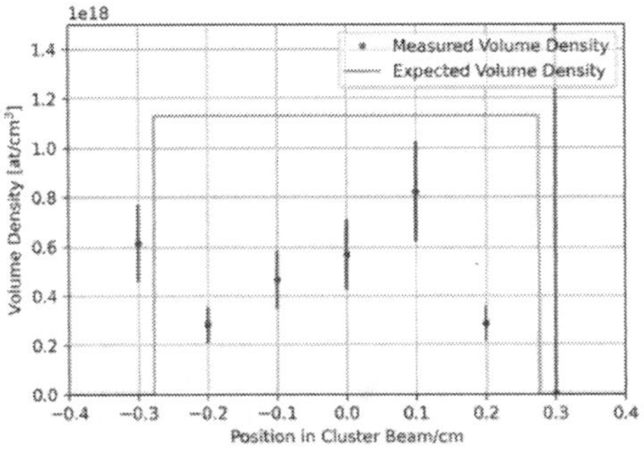

Fig. 5. Volume density from the shadowgraphy analysis shown in blue and expected volume density calculated from the gas flow through the nozzle in orange (work in progress).

However, two important results can be highlighted from fig. 5. First, it is evident that not all the gas that went initially through the nozzle is embedded in the large clusters. This result confirms the assumption made earlier that there are still smaller clusters that are below the resolution limit of the shadowgraphy images. Then, spatial structures can be recognized in the measured volume density. This means that the cluster beam is not homogeneously distributed, but that there are regions in the beam that are denser than others. This is in agreement with observations on high-intense hydrogen cluster beams, produced, e.g., at the PANDA cluster-jet target [10]. Further investigations related to this result will be carried out in the future.

3 Summary

With the shadowgraphy measurements and the subsequent analysis of the images, an important step has been made to investigate the properties of a hydrogen cluster-jet target.

When liquid hydrogen passes through a Laval nozzle, atomization takes place and clusters in the range between 3 µm and 10 µm are found to be produced. In comparison, no clusters are seen with gaseous hydrogen entering the nozzle, concluding that the formed clusters are then smaller than 3 µm.

Based on the analysis and the comparison of the integrals of the curves in fig. 5, it is seen that not all of the hydrogen is contained in large clusters, but there must also be smaller clusters, which are not detected on the images. The clusters found are in the order of a few micrometers, and a structure can be detected in the cluster beam. An upper limit of the cluster sizes of 10 µm has been determined.

The results presented here open possibilities for future measurements and give important information for experiments with cluster-jet beams, as well as for their analyses and simulations.

We acknowledge the excellent cooperation with O. Willi, E. Aktan, T. Burg, and M. Cerchez (HHU Düsseldorf) and also the excellent work done by our mechanical and electronic workshop. The research project was supported by BMBF (05P21PMFP1) and the EU's Horizon 2020 programme (824093).

References

1. P̄ANDA Collaboration, Technical Design Report for the P̄ANDA Internal Targets: The Cluster-Jet Target and Developments for the Pellet Target. Technical report. FAIR. panda.gsi.de/system/files/user_uploads/u.kurilla/RE-TDR-2012-002.pdf (2012).
2. S. Grieser, Doctoral thesis, Westfälische Wilhelms-Universität Münster (2018).
3. S. Grieser et al., Rev. Sci. Instrum. **90**, 043301 (2019).
4. L. Leßmann, Doctoral thesis. Westfälische Wilhelms-Universität Münster (2021).
5. B. Aurand et al., Phys. Plasmas **26**, 073102 (2019).
6. M. Cerchez et al., High Power Laser Sci. Eng. **7**, e37 (2019).
7. K. Oum et al., Phys. Chem. Chem. Phys. **5**, 5467-5471 (2003).
8. L. Halstenberg, Bachelor's thesis, Westfälische-Wilhelms-Universität Münster (2023).
9. A. Täschner, E. Köhler, H.-W. Ortjohann, A. Khoukaz, J. Chem. Phys. **139**, 234312 (2012).
10. S. Grieser, Master's thesis, Westfälische Wilhelms-Universität Münster (2014).

 https://doi.org/10.1051/epjconf/202328502002

The Münster cluster-jet target for the future $\overline{\text{P}}$ANDA experiment

Sophia Vestrick[1*], *Philipp* Brand[1], *Daniel* Bonaventura[1], *Hanna* Eick[1], *Christian* Mannweiler[1], and *Alfons* Khoukaz[1]

[1]Westfälische Wilhelms-Universität Münster, Institut für Kernphysik, Wilhelm-Klemm-Straße 9, 48149 Münster, Germany

Abstract. For high precision storage ring experiments as the future $\overline{\text{P}}$ANDA experiment, very sophisticated internal targets have to be used. For this purpose, a state-of-the-art cluster-jet target was developed at the University Münster. Basically, hydrogen is cooled to cryogenic temperatures and pressed through a specially shaped Laval nozzle to form a cluster-jet expanding into vacuum. Due to the stability and large mass of the clusters, a practically undisturbed flight path in vacuum of above 5 m is possible, leading to manifold possible applications, including the interaction with a storage ring beam at a distance of 2.25 m as desired for the $\overline{\text{P}}$ANDA experiment. With a first prototype target, the "proof-of-principle" was delivered, and after first improvements the world record in target thickness in such large distance to the nozzle was measured. Based on this work, the final $\overline{\text{P}}$ANDA cluster-jet target was developed and built up, and is presented in this article.

1 Introduction

The $\overline{\text{P}}$ANDA experiment (antiProton ANihilation at Darmstadt) will investigate antiproton proton collisions with the utmost precision at the antiproton high energy storage ring (HESR) at the facility for antiproton and ion research (FAIR), which is currently under construction. It aims to address fundamental questions of the quantum chromodynamics, i.e., the theory of strong interaction and hadron physics [1,2]. The antiprotons with a beam momentum of up to 15 $\mathrm{GeV/c}$ will interact with protons delivered by a cluster-jet target. Due to the 4π detector surrounding the interaction point (IP), the cluster source and target beam dump each need to be located at a distance of 2.25 m and 2.35 m to the IP, respectively, connected by narrow jet beam pipes. For the success of this experiment, it is indispensable to achieve world record target thicknesses in such large distances to the nozzle and to develop a new state-of-the-art cluster-jet target.

A first prototype cluster-jet target was set up in Münster. Here, first investigations on beam properties were performed and crucial improvements were introduced [3,4]. Importantly, the maximal useful target thickness [5] of $4.5 \cdot 10^{15}$ $\mathrm{atoms/cm^2}$ at the IP was reached and even exceeded, here in a distance to the nozzle of 2.1 m, with a target thickness of $(4.85 \pm 0.24) \cdot 10^{15} \mathrm{atoms/cm^2}$ [6].

Based on the obtained insights into the production of cluster-jets and setup handling, the final Münster Cluster-Jet Target was designed and built up, which is the topic of this article.

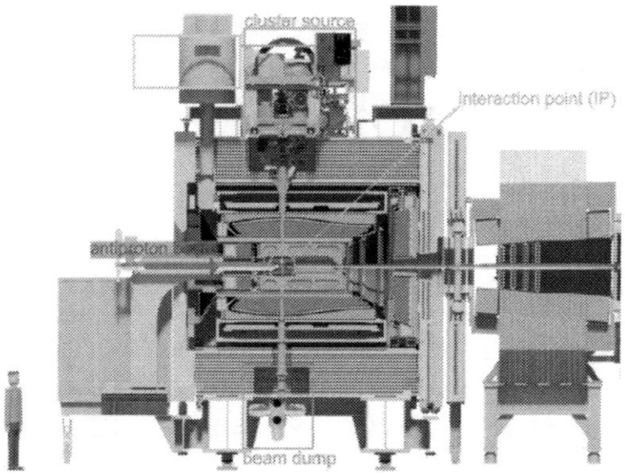

Fig. 1. Cross-section of the $\overline{\text{P}}$ANDA detector and the integrated $\overline{\text{P}}$ANDA cluster-jet target.

2 Münster Cluster-Jet Target for the $\overline{\text{P}}$ANDA experiment

The cluster-jet target can be separated into several subcomponents (fig. 1): A cluster-jet source, which will be placed on top of the detector frame, will generate small droplets of hydrogen in the nm to µm scale, also referred to as clusters. These clusters will need to travel for 2.25 m as a cluster-jet through a narrow target vacuum beam pipe to the interaction point. There, they are going to interact with the antiprotons of the stored beam. Afterwards, all clusters that did not interact need to be dumped in the differentially pumped beam dump, which will be placed at a distance of 2.35 m to the interaction point.

The cluster-jet source and a first version of the beam dump are currently in operation at the Cooler Synchrotron COSY in Jülich, where first insights on the

* Corresponding author: s_vest01@uni-muenster.de

interaction of clusters with a stored proton beam were already obtained.

2.1 Cluster-Jet Source

Within the cluster-jet source (fig. 2), hydrogen is cooled down by a two-stage cold head to temperatures ranging from 20 to 40 K and pressed with typically 5 to 20 bar through a convergent-divergent Laval nozzle expanding into a first vacuum chamber. Due to vacuum conditions better than 10^{-1} mbar, scattering with residual gas background is negligible. The resulting cluster-jet is separated from the gas background by a skimmer and afterwards tailored by a collimator, each optimizing the vacuum conditions of the differentially pumped stages. After a transition vacuum chamber, with a vacuum better than 10^{-4} mbar, the cluster-jet is injected into the accelerator vacuum.

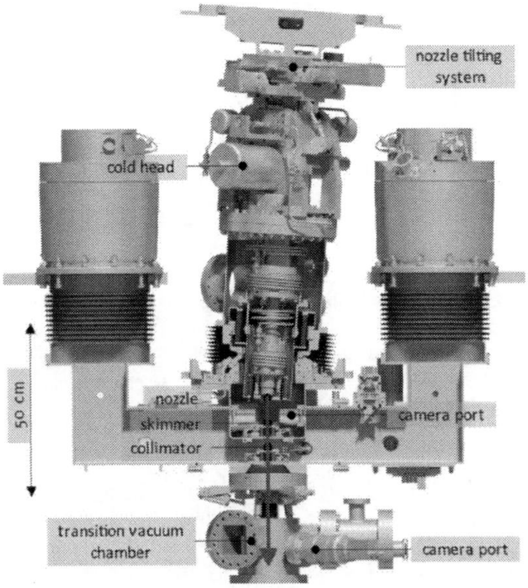

Fig. 2. Cluster-jet source. Cooled hydrogen is pressed through a Laval nozzle resulting in a defined cluster-jet propagating from top to bottom.

2.1.1 Convergent-divergent Laval Nozzle and Cluster Production Process

The convergent-divergent shape of Laval nozzles with narrowest diameters of approximately 30 µm leads to the possibility of forming hydrogen clusters, but is also a challenge in producing them.

Since the inner geometry needs to fulfil several special requirements, the desired Laval nozzles are not purchasable but need to be fabricated inhouse. Hence, a negative of the desired outlet is lathed out of aluminium. After a thin silver coating as separating layer, the negative is galvanized in a sulfuric acid copper bath by applying a DC current. As a sufficient diameter of copper is reached, the outer geometry is lathed and the convergent inlet is drilled. The most challenging part is now to drill the short connection channel between inlet and outlet of the nozzle as the resulting narrowest diameter is only 30 µm. Hence, this is a multi-step drilling process. Afterwards the aluminium outlet negative is removed, and the final nozzle is obtained.

Within these nozzles, two different cluster production mechanisms are possible, depending on the state of the fluid upstream of the nozzle.

In general, the hydrogen is accelerated to the narrowest nozzle diameter due to the convergence of the nozzle inlet until it reaches speed of sound [7]. In the subsequent divergent outlet of the nozzle, assuming gaseous hydrogen, the atoms are further accelerated [8]. Due to adiabatic cooling, the relative momentum of the atoms is reduced leading to the formation of condensation nuclei by means of three-body interactions. Further impacts with the surrounding co-flowing cryogenic gas lead to a growth of the clusters until millions of atoms are accumulated in single clusters [7, 8]. Assuming, on the other hand, liquid hydrogen being in front of the nozzle, the atoms are also accelerated in the divergent outlet. Here, at appropriate operation conditions, i.e., for typical pressures between 5 and 20 bar and temperatures $\gtrsim 20$ K, the liquid hydrogen atomizes and forms a spray [9].

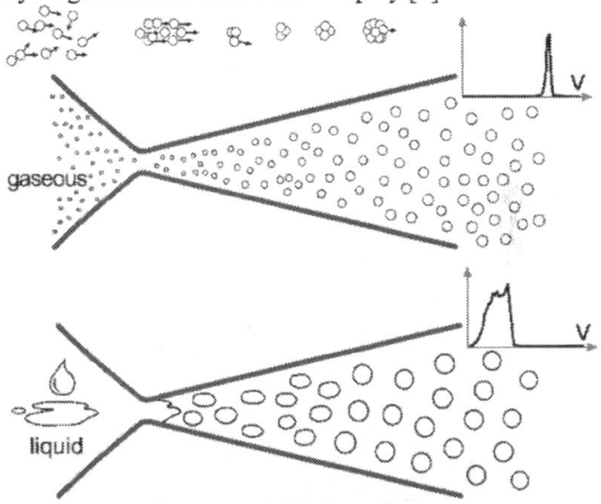

Fig. 3. Cluster production processes. *Top:* Clusters are formed from gaseous hydrogen by three-body interactions. *Bottom:* Clusters are formed from liquid hydrogen by an atomization process. Clusters formed from liquid hydrogen are larger and slower than clusters formed from gaseous hydrogen. (picture taken from [10])

Furthermore, when operating the cluster-jet target with liquid hydrogen, depending on the operational conditions, the resulting cluster-jet beam exposes high intense regions, referred to as core beams. Extracting these core beams allows for the desired high target thicknesses [6,11] of routinely 10^{15} atoms/cm^2. With different operation conditions smaller thicknesses are accessible. With gaseous hydrogen being in front of the nozzle, even thicknesses down to 10^{13} atoms/cm^2 or less are possible.

2.1.2 Movable Skimmer and Collimator and Nozzle Tilting System

In order to separate the most intense core beam of the produced cluster-jet from its surrounding residual gas jet, a conical orifice, called skimmer, with an opening diameter of 0.7 mm allows for the jet to enter the next vacuum chamber. A second conical orifice, called collimator, with an opening of $800 \times 220\ \mu m^2$ tailors the cluster-jet to its final rectangular shape further reducing the residual gas background.

After passing a transition vacuum chamber, the cluster-jet enters the target beam pipe. Over a distance of nearly 2 m, the pipe diameter is gradually reduced to 2 cm at the interaction point. In total, the most intense core beam of the cluster-jet needs to be aligned through nearly 5 m of the beam pipe to enter the beam dump and to reach highest target thicknesses at the interaction point. This can be achieved by tilting the nozzle and moving both the skimmer and collimator, which are each mounted on a stepper motor driven cross table.

2.2 Cluster-Jet Beam Dump

After passing the interaction point with the circulating antiproton beam, the remaining cluster-jet needs to be pumped away in a beam dump. Based on a prototype beam dump currently in operation, the final beam dump was developed. This beam dump will be located 2.35 m below the interaction point and will consist of three differentially pumped beam dump stages, which will be separated by adjustable orifices allowing to minimize the gas backflow (fig. 4). Additionally, within the beam dump several diagnostic tools will be implemented. These tools already have been tested and are described in the following.

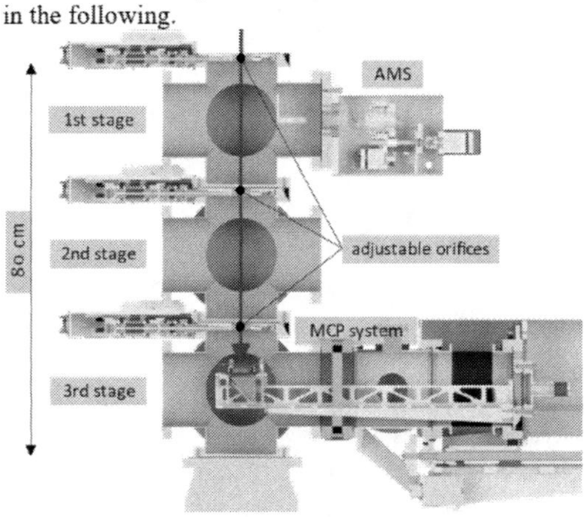

Fig. 4. Cluster-jet beam dump. After crossing the antiproton beam the cluster-jet will be dumped in the differentially pumped beam dump allowing for minimal gas backflow. Several beam diagnostic tools will be implemented.

2.2.1 *Absolute Thickness Monitor System (AMS)*

Within one of the beam dump stages, an absolute thickness monitor system will be installed. Here, two orthogonally oriented rods are moved one after another stepwise through the cluster-jet. Impinging clusters lead to a pressure increase proportional to the density of the cluster-jet. The pressure is measured depending on the rod position, which is shown in fig. 5. With the amplitude of the pressure profile normalized to the overlap of the moving rod and the cluster-jet $p_{0/A}$, the pumping speed S, the Avogadro constant N_A, the mean cluster velocity v, the universal gas constant R and the ambient temperature T, the density ρ of the cluster-jet is determinable by

$$\rho_H = 2 \cdot \frac{p_{0/A} S N_A}{v R T} \quad (1)$$

and with this and the dimension of the cluster-jet also its thickness.

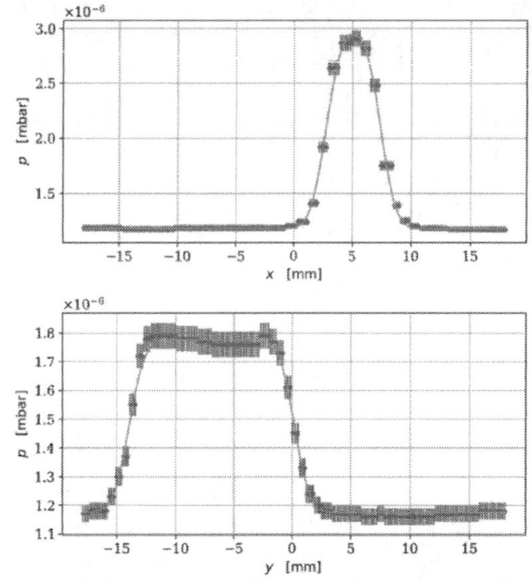

Fig. 5. Typical pressure profiles measured with the AMS in x- (*top*) and y-direction (*bottom*). From these profiles, the cluster-jet width in both directions and the actual position are directly observable. With further calculations (see eq. 1) also the cluster-jet density and the thickness are obtainable. It is to note that due to the rectangular collimator, also the cluster-jet has a rectangular cross-section with the long axis in particle beam direction (here y-direction) to maximize their overlap.

2.2.2 *Multi-Channel Plate (MCP) System*

Within the lowermost beam dump stage, an MCP system will be able to be moved into the cluster-jet allowing for several diagnostic analyses. It consists of a two-staged MCP in chevron assembly, a phosphor screen and a camera [12]. While taking data during the beam-target interaction at $\overline{\text{P}}$ANDA, the MCP system can be moved out of the cluster-jet, allowing for optimal vacuum conditions. Due to the small kinetic energy of neutral clusters and thus the negligible detection efficiency of the MCP, this system can only detect ionized clusters, which are accessible in two different ways.

In beam-target interactions during $\overline{\text{P}}$ANDA data taking operation, the antiproton beam only ionizes clusters in the vertex zone. These clusters hit the MCP and the subsequent electron avalanche lets the phosphor screen light up, which is detectable by the camera. Thus, an online vertex zone visualization with its two-dimensional size distribution is possible [12]. Moreover, with several calibration measurements it is possible to monitor the luminosity online during the $\overline{\text{P}}$ANDA experimental runs.

In combination with an electron gun, the total cluster-jet is ionized and thus visualized onto the phosphor screen. This is a useful tool in aligning the cluster-jet through its narrow beam pipe and is, after calibration measurements with the AMS, a faster way of determining the cluster-jet thickness [13].

Furthermore, the electron gun can be operated in a pulsed mode. Within the time window for ionization of 20 µs, a single cluster is ionized and thus the start signal for time-of-flight measurements is known. As this cluster hits the MCP system, the electron avalanche forms a voltage pulse at the phosphor screen, which can be measured and used as stop signal [13]. The time difference can then be used for a precise velocity measurement of single clusters leading to the cluster velocity distribution. From this, the mean velocity can be determined [7], which is needed for the thickness determination with the AMS as well as with the MCP system.

3 Summary

With the P̄ANDA cluster-jet target, a state-of-the-art cluster-jet target will be available for a broad variety of experiments at the future HESR. In order to reach the desired high target thickness, liquified, pressurized hydrogen is fed through a specially designed and manufactured Laval nozzle. This results in a high-intense cluster-jet, which is extracted and shaped by a movable skimmer and collimator. The cluster-jet is guided over nearly 5 m through a target beam pipe, which has a width of down to 2 cm at the interaction point, until it is dumped in a differentially pumped beam dump. At the interaction point in 2.25 m distance to the nozzle, a target beam with a maximum thickness of up to $4.5 \cdot 10^{15}\ \mathrm{atoms/cm^2}$ is placed in a high vacuum, with adjustable thickness over several orders of magnitude. Within the beam dump numerous self-developed diagnostic tools allow for determination of many properties as, e.g., cluster-jet position, cluster velocity, target thickness, target density, beam-target-interaction region, and luminosity. The delivered target is windowless and without any time structures. Furthermore, the target material is modifiable over a wide range of elements and compounds.

We want to acknowledge the support of the electrical and mechanical workshops in our institute for contributing to the final setup and the many steps in-between. This work was supported by BMBF, GSI F&E and HORIZON2020.

References

1. P̄ANDA Collaboration et al., *Strong Interaction Studies with Antiprotons*, arXiv: 0903.3905 (2009).
2. G. Barucca et al., Eur. J. Phys **A57**, 44, DOI: 10.1140/epja/s10050-021-00475-y (2021).
3. A. Täschner, Doctoral Thesis, Westfälische Wilhelms-Universität Münster (2012).
4. E. Köhler, Doctoral Thesis, Westfälische Wilhelms-Universität Münster (2015).
5. P̄ANDA Collaboration, *Technical Design Report for the P̄ANDA Internal Targets: The Cluster-Jet Target and Developments for the Pellet Target.* Technical report. FAIR. panda.gsi.de/system/files/user_uploads/u.kurilla/RE-TDR-2012-002.pdf (2012).
6. S. Grieser, Doctoral Thesis, Westfälische Wilhelms-Universität Münster (2018).
7. A. Täschner, E. Köhler, H.-W. Ortjohann, A. Khoukaz, J. Chem. Phys. **139**, DOI: 10.1063/1.4848720 (2013).
8. H. Pauly, Atom, Molecule, and Cluster Beams 1, Springer (2000).
9. C.C. Miesse, Ind. Eng. Chem. **47 (9)**, 1690–1701, DOI: 10.1021/ie50549a013 (1955).
10. S. Vestrick et al., J. Supercrit. Fluids **188**, DOI: 10.1016/j.supflu.2022.105686 (2022).
11. B. Hetz, Masters Thesis, Westfälische Wilhelms-Universität Münster (2017).
12. A. Khoukaz et al., Nucl. Instrum. Methods Phys. Res. **A735**, 12, DOI: 10.1016/j.nima.2013.08.085 (2014).
13. S. Vestrick, Masters Thesis, Westfälische Wilhelms-Universität Münster (2020).

https://doi.org/10.1051/epjconf/202328503001

The process of molecular plating and the characteristics of the produced thin films – What we have learned in 60 years and what is still unknown

Ernst Artes[1, 2, 3*], *Christoph E.* Düllmann[1, 2, 3], *Carl-Christian* Meyer[1, 2], and *Dennis* Renisch[1,2]

[1]Johannes Gutenberg Universität Mainz, 55099 Mainz, Germany
[2]Helmholtz-Institut Mainz, 55099 Mainz, Germany
[3]GSI Helmholtzzentrum für Schwerionenforschung GmbH, 64291 Darmstadt, Germany

Abstract. Molecular plating is a well-established and widely used method for producing thin films of various elements, which are used in variety of nuclear physics applications. Sixty years have passed since the method was established, and some insights into the chemical process underlying the method and the composition of the thin films have been gained. A brief overview of what has been learned about molecular plating since its introduction and the methods applied in the characterization of molecular plated thin films is given here. Through various spectroscopic and microscopic methods, the process of molecular plating and the chemical composition are gradually being elucidated, albeit we still do not understand all aspects.

1 Introduction

The molecular plating (MP) method for the production of mechanically stable thin films of radioisotopes was introduced by Parker and Falk in 1962 [1]. In this method, a salt of the respective element was dissolved in a small volume of diluted acids, added to an organic solution, e.g., alcohols or acetone and then electrochemically deposited. Nowadays, a constant current density of <1 mA/cm^2 over 1-2 h is applied. This results in high voltages of several hundred volts. MP produces uniform thin films with an even distribution of the deposited element. Such layers found widespread applications, e.g., as targets in ion beam experiments, as recoil sources, or as α-particle sources [3-12]. One example is the synthesis of superheavy elements (SHE), which can be produced in the nuclear fusion of a light projectile with a heavy target nuclide. All known isotopes of the elements with atomic number $Z \geq 113$ resulted from actinide-target based reactions [11]. Typical isotopes like ^{244}Pu, ^{248}Cm or ^{249}Bk are only available in small quantities, rendering production methods with high yields necessary. Therefore, MP has become the standard method for producing these targets [3,11,12]. For this application, the method works reliable up to film thicknesses of ≈ 800 μg/cm^2, with yields of usually over 90 % [2,3,4]. While thicker films can be produced for various elements, the method becomes unreliable for actinide layers. Consequently, actinide films that are substantially thicker have not been reported. Although molecular plating was introduced in 1962 [1] and has been widely used since, neither the mechanism of the deposition process, nor the exact composition of the resulting thin layer is fully understood. Parker and Falk coined the method "molecular plating" because they assumed at the time that the provided salts, e.g., the nitrates or chlorides, would be deposited as thin films [1]. Nowadays, it is well known that this is not the case. Many different analytical methods [2,12-18] have shown that the chemical species of the deposited element in the thin film differs from the species in the solution. Over time, mostly microscopic and also some spectroscopic methods have been used to analyse the thin films produced by MP and to understand how the process works, see e.g., [2,4,18,19]. The elucidation of the process and the layer properties appear necessary to produce improved films, which e.g., exceed the limitations in the layer thicknesses, as it appears desirable for, e.g., SHE production [4].

2 Methods

Many methods focused on the film's morphology, which led to the use of several microscopic methods, such as scanning electron microscopy (SEM) [6,17,21,22] and atomic force microscopy (AFM) [2,16,17]. Radiographic imaging (RI) [18,23,24] exploits the radioactive character of actinides to provide information on the spatial distribution of the radioisotope. Alpha and gamma spectroscopy [8,9,21,23,24] provide further quantitative information.

Neutron activation analysis (NAA) [25] can be used to activate inactive samples or long-lived isotopes,

* Corresponding author: erartes@uni-mainz.de

making gamma measurements possible that reveal the amount of deposited material.

With SEM, energy-dispersive X-ray spectroscopy (EDS) was also employed early on to determine the elemental composition of the thin films. [13,15,19,26]. However, these methods are all unable to provide information on the chemical composition of the thin films, which is why spectroscopic methods, like IR, Raman, X-ray diffraction (XRD), X-ray photoelectron spectroscopy (XPS), or ion beam spectroscopy like elastic recoil detection analysis (ERDA), Rutherford backscattering (RBS) or proton-induced X-ray emission (PIXE), were also used later [2,13,15,16,19,27-29].

Raman and IR measurements were made early on [15,19], but not recently pursued further. These methods allow determining the chemical composition. Both methods are complementary as the vibrations of the molecules are either IR or Raman active. Further information can be gained by XPS [2,16,17], which provides information on the chemical environment of the element. XRD measurements were used to investigate the crystalline composition of the films [13,14] to reach conclusions about the crystalline structure.

Most recently, various ion beam analyses have also been applied to molecular plated thin films. One such method is ERDA [4], which is particularly well suited for detecting light elements from lithium up to chlorine in the thin film. Complementary to this, RBS was also used [12,28,29] to allow drawing conclusions about the elemental composition of the films.

PIXE was also measured in combination with RBS [4]. In PIXE, a proton beam induces X-rays in the sample to obtain further information about the film. It can also be used as an imaging method, which allows both microscopic and spectroscopic information to be obtained simultaneously.

3 Elucidation of the MP process

In 1959, even before MP was introduced, Hansen [30] had published a theory on electrochemical deposition from water or mixtures of water with alcohols or acetone. He suggested that water is reduced in the process, producing hydroxide ions that lead to basic precipitation of the actinides [19,30]:

$$2\ H_2O + 2\ e^- \rightarrow H_2 + 2\ OH^- \tag{1}$$

In 1975, Ramaniah et al. [15] were able to determine through thermogravimetric, EDS and IR analyses that no nitrates were present in MP films. In particular, the IR analyses of uranium films showed a U-O stretching vibration of UO_2^{2+}, which led to the assumption that uranium was deposited as an oxide or hydroxide. In addition, by adding phenolphthalein, the presence of water was observed in the isopropyl alcoholic solution and a hydroxide-rich layer was observed at the cathode [15]. Sadi et al. [19] proposed the chemical composition to be $UO_2(OH)_2$, based on the atomic ratios determined by EDS measurements. They proposed the hydroxide ions to act as bridge ligands between the uranyl ions as a possible structure in the thin film. With more analytical investigations, it became clear that the MP process is even more complex than a hydroxide precipitation, and the thin films contain more than only oxide and hydroxide species [2,14-17,19,27].

4 Elucidation of the properties of MP films

With the help of microscopic methods, the thin layers could be examined in detail. Since the focus was particularly on the deposition of uranium and plutonium in the beginning, the activity of these elements was used to take radiographic images of the samples [18]. These images showed that the activity of the films was particularly homogeneously distributed, which is visible in Fig. 1.

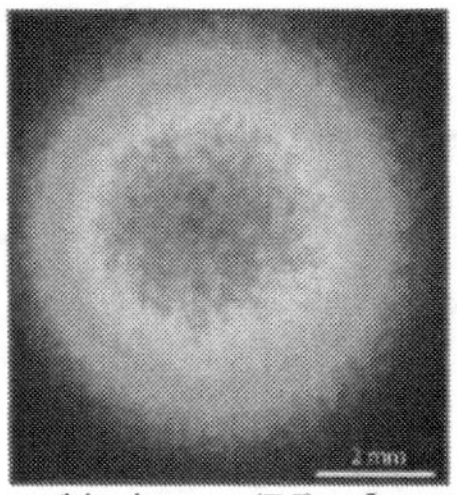

Fig. 1. A radiographic image (RI) of a terbium thin film, produced by MP, with a layer thickness of 500(50) μg/cm^2. Natural terbium was neutron activated in the TRIGA Mainz research reactor [25] before plating.

Radiographic images also made it possible to identify the influence of various experimental parameters, such as cell design or electrode shape, on the deposition. In this way, information could be obtained about the influence of the geometry of the anode, i.e., the counter-electrode, on the distribution of the activity in the thin film [23,24,31,32]. Initial experiments also show that stirring with a rotating anode or an ultrasonic stirrer during deposition can lead to shorter deposition times by forcing convection in the solution. However, the choice of parameters is crucial [7,23,24]. Due to the activity of uranium and thorium isotopes, alpha and gamma spectra of samples produced by molecular plating could also be taken, so that the deposition yield could be optimized by adjusting the deposition time and the applied current [1-3,18-23,31,32].

The usage of a SEM with a spatial resolution of <1 μm [4] made it clear that the molecular plated thin films were not homogeneous and closed surfaces as was inferred from radiographic images, which have a spatial resolution of >100 μm [33], but consisted of many cracks and tiles, referred to as "mudcracking" [1,4,20,26,27]. Pictures of such tiles are visible in Fig. 2. Today, SEM measurements are a standard method to characterise molecular-plated thin films in higher resolution than a usual light microscope [4,16,18,21].

Later Liebe et al. [13] confirmed the absence of nitrogen, as first reported by Ramaniah et al. [15], in the thin MP films by EDS measurements. Therefore, nitrates cannot be present in the thin films, which

 https://doi.org/10.1051/epjconf/202328503001

definitely refuted Parker and Falk's theory of salt re-deposition [1].

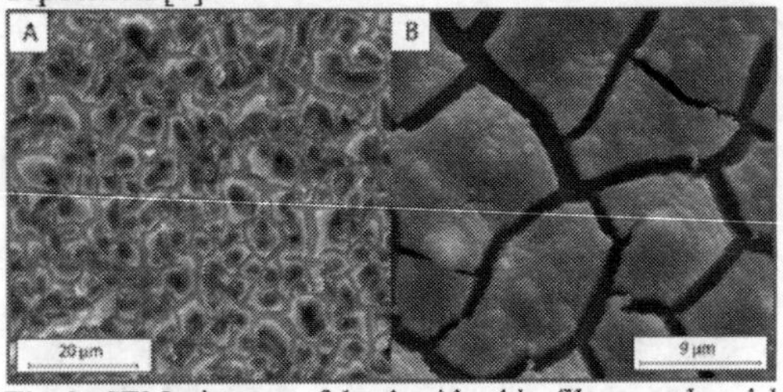

Fig 2. SEM pictures of lanthanide thin films produced by molecular plating. A) Erbium thin film. B) Silver sputtered lanthanum thin film. Both thin films have a layer thickness of 500(50) µg/cm^2.

Vascon et al. [16] were able to show by AFM measurements in solution that cracks are not present in the native film during the MP process, but form only after removal of the solvent, during drying after deposition. Fig. 3 shows an exemplary AFM image of a Tb thin film, after the drying process. They could also show that different solvents impact the quality of the thin film [16] and explained this by the fact that different solvents have different boiling points and different vapour pressures. It is assumed that residues of the solvents are present in the film after deposition. Evaporation during drying leads to rupturing of the films. They were able to observe that deposition from DMF (boiling point: 153 °C), leads to less mudcracking than deposition from isobutanol (boiling point: 108 °C).

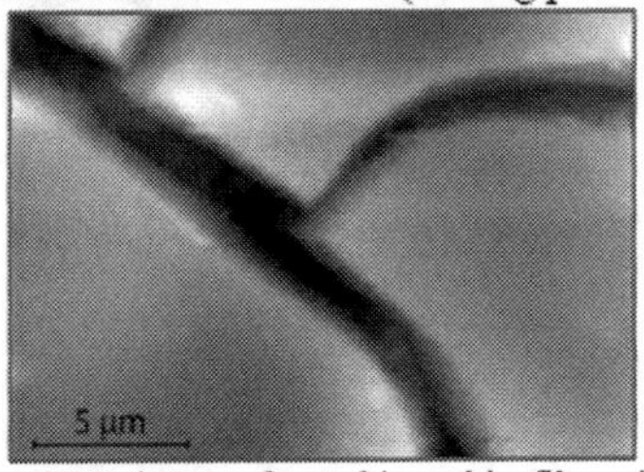

Fig 3. An AFM picture of a terbium thin film with a layer thickness of 100(10) µg/cm^2, which was produced by molecular plating.

Vascon et al. [2,16,17] also found out with the help of XPS measurements that the produced thin films not only contained oxides and hydroxides, but also organic components. Besides adventitious carbon contamination, aliphatic and aromatic carbon was found [2,16,17]. It did not disappear even after sputtering with 4 kV Ar^+-ions for 10 min. They thus assumed that the carbon was a constituent of the thin film. They attributed the signals to the presence of residue solvents, carbonates and/or carboxylates, which was assumed to form by the electrolysis of the used solvents. Further XPS measurements also showed that the lanthanide is not reduced in the process of plating but remains in the +III oxidation state [2,13,15-17].

In 2016, Choi and Chung [13] did XRD measurements on molecular-plated thin films to understand what happens to the thin films when they are calcinated. They identified, next to oxides and hydroxides, also predominantly oxycarbonates after a short calcination time, further proving that significant amounts of organic molecules were present in the molecular plated thin films.

Mayorov et al. [26] investigated Gd MP thin films after irradiation with ion beams ($^{48}Ca^{7+}$ with 3.8 MeV/u and 4.5 x 10^{14} ions/cm^2 and $^{45}Sc^{6+}$ with 4.7 MeV/u and 4.0 x 10^{15} ions/cm^2). In the XRD measurements, the only visible compound before and after the irradiation was Gd_2O_3. The EDS spectra showed, in addition to gadolinium, also carbon which they attributed to carboxylates because of this prior assignment by Vascon et al. [2,16,17]. Mayorov et al. assumed that organic residues evaporated during the irradiation leading to a decreased carbon signal. The SEM picture showed that the typical cracked film became smoother and thinner after the irradiation, closing the cracks, which were visible before [26].

Although much has been learned about the MP method and the produced thin films, much is still unknown. It is not still clear why high voltages of several hundred volts are necessary for this method to work sufficiently well. It is also unknown what happens to most of the current that is applied during the MP process, and whether this loss of charge has any influence [2]. It is also unknown why the layer thickness is limited to 800 µg/cm^2 and how this limit can be exceeded [3,4]. Although individual components such as oxides, hydroxides, carbonates, and carboxylates have been identified in the thin films [2,7,12-17,19,27], the ratios of the individual components to each other are still unknown. It is therefore not yet possible to determine a sum formula for the composition of the film.

5 Conclusion

Since the establishment of molecular plating by Parker and Falk [1] 60 years ago, knowledge about the method has been gained via many different methods. Via different microscopic methods, like RI, SEM and AFM, and different spectroscopic methods, like alpha and gamma spectroscopy, EDS, XPS, XRD, RBS, IR, and Raman spectroscopy [2,4,7,13-17,19,27], a more precise understanding of the process of MP and of the chemical composition of the thin films was obtained.
It became clear that no nitrates or chlorides are deposited, but that the thin film consisted mainly of oxides and hydroxides. In addition, carboxylates and carbonates are also present [2,16,17]. It was also shown that the choice of solvent during deposition has an influence on the film quality [16].

By exposing MP films to heavy ion beams, they are further modified, making them thinner and closing the cracks [26].

We still do not fully know, what the stochiometric composition of MP films is and what influences the success or failure of a deposition [4,16,21,22]. In addition, it is still not understood, why the thicknesses of the produced thin films are limited and how to overcome this obstacle [3,4].

Therefore, further studies of the MP process and of the resulting layers are necessary, alongside further development to improve the method for future applications. In this way, further systematic studies could be carried out to understand the influence of CO_2 and water on the MP process.

We acknowledge funding from the German Federal Ministry for Research and Education (project 05P21UMFN2). The contribution of Sebastian Herz (JGU Mainz) is appreciated.

References

1. W. Parker, R. Falk, Nucl. Instr. And Meth. **16**, 355-357 (1962)
2. A. Vascon, S. Santi, A. A. Isse, T. Reich, J. Drebert, H. Christ, Ch. E. Düllmann, K. Eberhardt, Nucl. Instrum. Methods. Phys. Res. A, **696**, 163-175 (2012)
3. K. Eberhardt, W. Brüchle, Ch. E. Düllmann, K. E. Gregorich, W. Hartmann, A. Hübner, E. Jäger, B. Kindler, J. V. Kratz, D. Liebe, B. Lommel, H.-J. Maier, M. Schädel, B. Schausten, E. Schimpf, A. Semchenkov, J. Steiner, J. Szerypo, P. Thörle, A. Türler, A. Yakushev, Nucl. Instrum. Methods. Phys. Res. A, **590**, 134-140 (2008)
4. Ch. E. Düllmann, E. Artes, A. Dragoun, R. Haas, E. Jäger, B. Kindler, B. Lommel, K.-M. Mangold, C.-C. Meyer, C. Mokry, F. Munnik, M. Rapps, D. Renisch, J. Runke, A. Seibert, M. Stöckl, P. Thörle-Pospiech, C. Trautmann, N. Trautmann, A. Yakushev, J. Radioanal. Nucl. Chem. (2022) https://doi.org/10.1007/s10967-022-08631-4
5. J. Runke, Ch. E. Düllmann, K. Eberhardt, P. A. Ellison, K. E. Gregorich, S. Hofmann, E. Jäger, B. Kindler, J. V. Kratz, J. Krier, B. Lommel, C. Mokry, H. Nitsche, J. B. Roberto, K. P. Rykaczewski, M. Schädel, P. Thörle-Pospiech, N. Trautmann, A. Yakushev, J. Radioanal. Nucl. Chem. **299**, 1081–1084 (2014)
6. B. Lommel, Ch. E. Düllmann, B. Kindler, D. Renisch, Eur. Phys. J. A, (accepted for publication)
7. W. Loveland, J. D. Baker, J. Radioanal. Nucl. Chem **282**, 361-363 (2009)
8. W. Loveland J. Radioanal. Nucl. Chem **307**, 1591-1594 (2016)
9. A. Vascon, Ch. E. Düllmann, K. Eberhardt, B. Kindler, B. Lommel, J. Runke, Nucl. Instrum. Methods. Phys. Res. A **655**, 72-79 (2011)
10. J. P. Greene, R. V. F. Janssens, I. Ahmad, Nucl. Instrum. Methods. Phys. Res. A **438**, 119-123 (1999)
11. R. Haas, T. Kieck, D. Budker, Ch. E. Düllmann, K. Groot-Berning, W. Li, D. Renisch, F. Schmidt-Kaler, F. Stopp, A. Viatkina, Hyperfine Interactions **241**, 25 (2020)
12. R. Haas, M. Hufnagel, R. Abrosimov, Ch. E. Düllmann, D. Krupp, C. Mokry, D. Renisch, J. Runke, U. W. Scherer, Radiochim. Acta **108**, 923-941 (2018)
13. D. Liebe, K. Eberhardt, W. Hartmann, T. Häger, A. Hübner, J. V. Kratz, B. Kindler, B. Lommel, p. Thörle, J. Steiner, Nucl. Instrum. Methods. Phys. Res. A, **590**, 145-150 (2008)
14. J. Choi, Y. H. Chung, Journal of Nanomaterials, **2016**, 5140219 (2016)
15. M. V. Ramaniah, R. J. Singh, S. K. Awasthi, S. Prakash, Int J Appl Radiat Isot **26**, 648-650 (1975)
16. A. Vascon, S. Santi, A. A. Isse, A. Kühnle, T. Reich, J. Drebert, K. Eberhardt, Ch. E. Düllmann Nucl. Instrum. Methods Phys. Res. A **714**, 163-175 (2013)
17. A. Vascon, J. Runke, N. Trautmann, B. Cremer, K. Eberhardt, Ch. E. Düllmann, Appl. Radiat. Isot. **95**, 36-43 (2015)
18. N. Getoff, H. Bildstein, Nucl. Instr. And Meth. **36**, 173-175 (1965)
19. S. Sadi, A. Paulenova, P. R. Watson, W. Loveland, Nucl. Instrum. Methods. Phys. Res. A, **655**, 80-84 (2011)
20. Ch. E. Düllmann, M. Block, F. P. Heßberger, J. Khuyagbaatar, B. Kindler, J. V. Kratz, B. Lommel, G. Münzenberg, V. Pershina, D. Renisch, M. Schädel, A. Yakushev, Radiochim. Acta, **110**, 417-439 (2022)
21. W. Parker, H. Bildstein, N. Getoff, Nucl. Instr. And Meth. **26**, 55-60 (1964)
22. W. Parker, H. Bildstein, N. Getoff, H. Fischer-Colbrie, H Regal, Nucl. Instr. And Meth. **26**, 61-65 (1963)
23. M. T. Crespo, Appl. Radiat. Isot. **70**, 210-215 (2012)
24. H. Klemenčič, L. Benedik, Appl. Radiat. Isot. **68**, 1247-1251 (2010)
25. K. Eberhardt, C. Geppert, Radiochim. Acta **107**, 535-546 (2019)
26. D. A. Mayorov, E. E. Tereshatov, T. A. Werke, M. M. Frey, C. M. Folden III, Nucl. Instrum. Methods. Phys. Res. B **407**, 256-264 (2017)
27. C.-C. Meyer, A. Dragoun, Ch. E. Düllmann, R. Haas, E. Jäger, B. Kindler, B. Lommel, A. Prosvetov, M. Rapps, D. Renisch, P. Simon, M. Tomut, C. Trautmann, A. Yakushev, Nucl. Instrum. Methods. Phys. Res. A, **1028**, 166365 (2022)
28. O. A. Wasson, R. A. Schrack, Nucl. Instrum. Methods. Phys. Res. A **282**, 194-198 (1989)
29. C. Stodel, EPJ Web of Conferences **229**, 02001 (2020)
30. P. G. Hansen, J. Inorg. Nucl. Chem. **12**, 30-37 (1959)
31. Young Gun Ko, J. Radioanal. Nucl. Chem. **326**, 861-877 (2020)
32. C. Ingelbrecht, A. Moens, R. Eykens, A. Dean, Nucl. Instrum. Methods. Phys. Res. A, **397**, 34-38 (1997)
33. R. Haas, L. Lohse, Ch. E. Düllmann, K. Eberhardt, C. Mokry, J. Runke, Nucl. Instrum. Methods. Phys. Res. A, **874**, 43-49(2017)

Spherical powders: control over the size and morphology of powders for additive manufacturing and enriched stable isotope nuclear targets

Michael Zach[1*], *Dane* Brashear[1], *Jonah* Duran[1], *Łukasz* Żrodowski[2], *Bartosz* Kalicki[2], *Tomasz* Choma[2], *Marcin* Sołowiow[2], and *Steven* Adler[3]

[1]Oak Ridge National Laboratory, Stable Isotope Materials and Chemistry, Oak Ridge, Tennessee, USA
[2]AMAZEMET | Warsaw University of Technology, 00-867 Warsaw, Poland
[3]A3DM Technologies, Sarasota, Florida, USA

Abstract. Metal powders are a fundamental starting point for fabricating many types of nuclear targets. Elemental powder properties can differ drastically between batches, even when using the same method. Therefore, the variation in morphology and the size of metal powders can cause variable quality and produce inconsistent results with what are otherwise proven target manufacturing techniques. Additive manufacturing has additional requirements for higher quality and more uniform feedstock. The production of spheroidized powders with uniform, reproducible properties and a narrow size distribution represents unexplored opportunities for experiments. These opportunities include experimenting with solid metals that can now flow like liquids, new options for powder handling and dispensing, and new target fabrication methods using additive manufacturing. The Stable Isotope Materials and Chemistry Group at Oak Ridge National Laboratory obtained an AMAZEMET rePowder ultrasonic metal atomization tool for creating limited batches of fully dense, free flowing, spherical powders with a narrow size distribution of extremely rare materials. Early results are presented with materials that were produced. The team explores the anticipated limits of this instrument with extremely rare materials (e.g., enriched stable isotopes) and highlights research into new fabrication techniques that provide additional options benefitting the international nuclear target community.

1 Introduction

The fabrication of materials, specifically enriched isotopes, into specific forms such as foils, discs, wires, and pellets is of great interest to the members of the International Nuclear Target Development Society (INTDS). The custom manufacturing of high precision targets with a production volume of one unit with exceptionally rare materials brings specific challenges. Although overall methods for producing targets are often well described, the practical task of fabricating targets can be problematic. Challenges include changes in personnel, infrequent fabrication runs, and the loss of familiar equipment. However, variations of powder may be most challenging. Because most elements are processed as a granular powder at some point during the production and transformation into a target, powders affect most targets. Slight variations in the synthesis of the powder or storage conditions could be the difference between success and failure. Each batch, and even variations within a batch, can make a large difference in the working properties [1].

2 Failures attributed to powder

Fig. 1. Commercially prepared 99.95% platinum.

Figure 1 shows a scanning electron microscope (SEM) image of a freshly opened batch of platinum powder from a commercial supplier. The size distribution ranges

* Corresponding author: ZachMP@ornl.gov

from sub micrometer to about 10 μm. Powders often further aggregate to form larger clusters that exceed 1 mm in size. The history of the powder from the time of manufacture and packaging at the factory to exposure to the environment in a laboratory can alter the powder. Furthermore, variations in humidity, O_2, or CO_2 levels can change the chemistry on the surface after it is opened. Fresh drying or hydrogen reduction cycles can achieve a more consistent starting powder material, but these processes do not address the effects of size. In addition, heating may help form particle clusters. Powders of different sizes naturally tend to segregate. The fine powders settle at the bottom of the original container, whereas the less dense, larger clusters of particles tend to float. Larger clusters are preferentially removed, and subsequent removals contain smaller particles until the end of the batch is used.

The effect of crystallinity and surface area is profound upon melting and casting. Natural ruthenium powder produced in the 1960s was pressed into briquettes and arc melted producing near-solid density castings (Fig. 2a). After purchasing of new material, the castings were filled with voids. Although the powders looked identical to the naked eye and at 100x magnification, higher magnification SEM images at 10kX magnification (Fig. 2d and 2e) revealed differences in the microstructure. The original powder was blocky crystals likely caused by slow precipitation from solution, whereas the new material was highly porous and as sponge-like, which originated from elevated temperature hydrogen reduction of $(NH_4)_3RuCl_6$ [2]. The extreme differences in surface area led to large amounts of trapped gas that would escape when the molten ruthenium cooled and caused a hollow horn like structure to push out of the surface (Fig. 2b and 2c). High-temperature vacuum sintering of pressed powder caused 4% shrinkage for the powder and 26% shrinkage for the sponge under the same conditions.

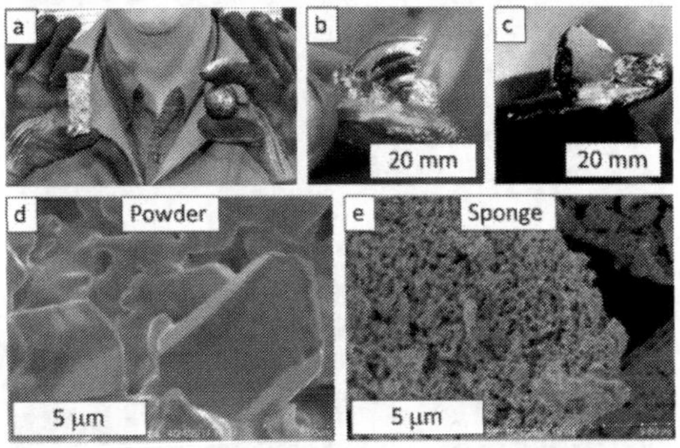

Fig. 2. Microstructural differences of ruthenium powder leading to a failure in casting. **a)** Completed high density ruthenium casting (left) and the intermediate melted button made from powder (right). **b) and c)** Button made from pressed sponge with hollow horn that grew out of button. **d)** Old-stock ruthenium powder that was successful. **e)** Newly obtained ruthenium sponge powder that led to failure.

Another common problem with powders is a direct result of the high surface area. When other elements are present—whether chemically bonded or physiosorbed onto the surface—they can yield fundamentally different properties. A recent example of this was in the processing of isotopically enriched ^{62}Ni pellets for the neutron capture production of ^{63}Ni [3]. In this example, pellets were successfully prepared for multiple production campaigns using a historically produced Oak Ridge calutron material. The process involves compression of ^{62}Ni powder within a steric acid lubricated die. The green pellets are further sintered to volatilize the steric acid and diffuse the grains together, which form dense hard pellets (Fig. 3a). When the first batch was consumed, a second batch of material was opened. Despite our well-established procedure, every pellet failed by melting and fusing itself to the tantalum tray during vacuum sintering (Fig. 3c). The powder-related failure resulted in delays and lost labor of the original pressing and sintering, recovery of the nickel from the tantalum tray (causing material loss and additional analytical costs), and lengthy troubleshooting. The initial thinking was a faulty thermocouple giving poor temperature control. Multiple trials were conducted that proved no issue with the furnace. SEM and energy dispersive spectroscopic (EDS) analysis was performed on a residual amount of the original material (Fig. 3b) to compare with the new material (circled in Fig. 3d) that showed phosphorous left from chemical separation. The trace amount of phosphorous was enough to form a eutectic with nickel causing melting at 200°C lower than the normal sintering temperature.

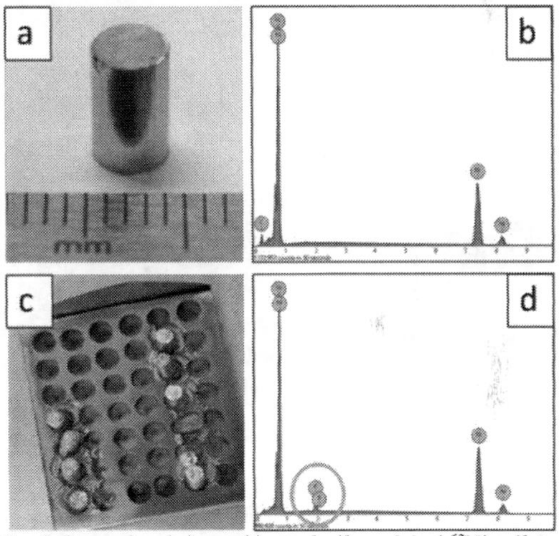

Fig. 3. Pressed and sintered isotopically enriched ^{62}Ni pellets.

3 More controllable uniform powder

As outlined in section 2, powders, when not properly understood or controlled, can lead to failures in otherwise routine processes. Powder production is fundamental to our targetry community, yet hard to gain control. This is complex because there are dozens of methods to make powders. Metal powders can be the result of direct production from ions, chemical reduction of precipitated powders, fracturing of ingots through mechanical attrition such as crushing [4] as shown with the $Bi_{0.6}Sb_{1.6}Te_3$ powder on the left side of Fig. 4, or a hydride/dehydride fracturing of select metals including most of the lanthanides. All these result in high surface area, with irregularly shaped particles ranging from nanoparticles to large chunks. The distribution of size can be narrowed by sieving, suspension in liquid, winnowed by moving gas, or by other mechanical means.

Spherical powders are preferred for some processes. Like miniature ball bearings, they will roll past one another and can be spread out as a continuous thin film when subjected to shear forces (Fig. 4, right) vs. a discontinuous film created by irregularly shaped powders from ball milling (Fig 4, left). These free flowing, well-characterized powders are the standard feedstock for additive manufacturing processes. Industrially, these can be produced in >50-ton batches using techniques such as gas or water atomization. These techniques intersect a stream of molten metal with a hypersonic jet of non-reactive gas, water, or steam. Although these industrially relevant metal powders (i.e., mostly alloys) are feedstocks for powdered metallurgy manufacturing, these same processes cannot be scaled to allow processing of small batches of specialty metals spanning the periodic table. Because of the large equipment required and the latency time to get steady state conditions before suitable powder is produced, most metal atomization is practical only with large volumes. The commercially available materials produced in the lowest quantity are typically precious metals for the jewelry industry. Silver, gold, and platinum alloys are typically produced in 10–100 kg batches. Losses for precious metals are acceptable only because of the large quantity per batch.

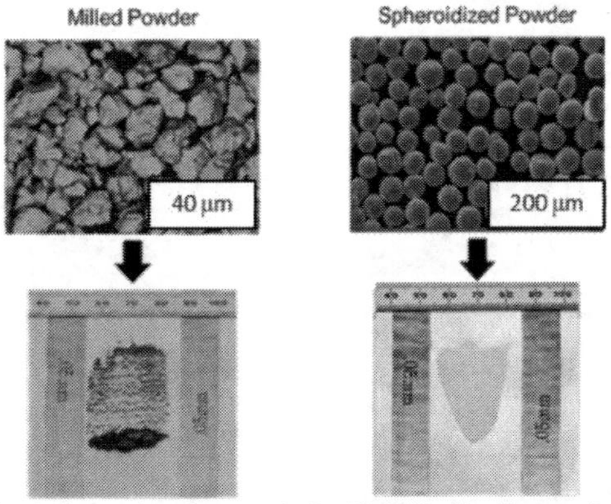

Fig. 4. Comparison between ball milled $Bi_{0.6}Sb_{1.6}Te_3$ powder on the left with commercially available spherical powder on the right (316 stainless steel). Photo: Fred List, III, (ORNL).

Compared with precious metals (3–5 cents per milligram), enriched stable isotopes can range from tens to more than a million times the value of precious metals. Ideally, conversion and recovery need to be high. Furthermore, empirical trials and errors must be minimized by selecting techniques that are both reproducible and material agnostic. Most elements that are not extremely reactive, toxic, or radioactive should be capable of being made into spherical powders using the same technique.

The discussion herein will be limited to three methods: aerosol jetting, machining, and ultrasonic atomization. Aerosol jetting was accomplished in a graphite crucible, with an approximate 5–10 µm hole in the bottom, that was pressurized to 40 psi with argon gas while being heated using an oxy-propane torch. Spherical powders are collected in a water reservoir. Although it was successful for bismuth, gold, and silver (Fig. 5), the process had a low conversion rate average of <10% and is limited to materials that have low reactivity to air, water, and graphite. This limits use of this method to silver, gold, platinum, palladium, tin, and bismuth. Copper was noticeably oxidized and metals such as cadmium and lead could have environmental, health, and safety concerns, even when performed in a hood.

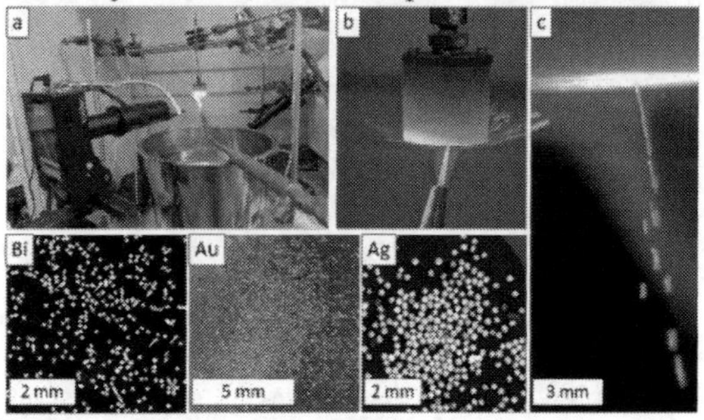

Fig. 5. Aerosol jetting. **5a** shows the overall experimental set up. **5b** and **5c** show details of torch heating and stream of metal being propelled out through a fine hole. Other images show samples of the metal powders.

CNC machining is a method that is universal to all metals but some need to be machined within an inert argon glovebox because of rapid oxidation. The first image of Fig. 6 shows a single particle of zinc being ejected off a 10,000-rpm single flute tungsten carbide mill. By controlling the spindle speed, depth of cut, and feed rate, controllable particle sizes could be obtained. The method is slow for production but is automated. The particles were vacuumed using a centrifugal particle separator with >99% collection efficiency. A few metals (e.g., ruthenium, zirconium, and rhodium) were found to be exceptionally hard, but grinding using nickel-bonded abrasive diamond bits were found to be effective. Particle sizes with the abrasive technique were finer and more difficult to contain. Water was used as lubricant for these metals, and recovery was considerably complicated. This method would need further development of the technique to routinely achieve >60% recovery.

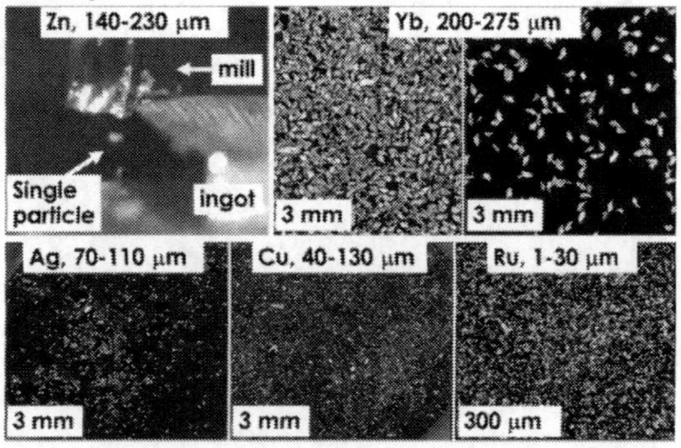

Fig. 6. Various metals produced by machining. High speed video frame shows a single particle being milled.

Ultrasonic atomization was the technique determined to be the most effective for our low quantities of metallic elements and alloy systems [6];

however, it required the most equipment and complexity for production. Fig. 7 shows a variety of materials that have been produced using this technique. The image in the upper left shows the instrument produced by AMAZEMET, a start-up from the Warsaw University of Technology. The instrument uses a grounded, water-cooled, ultrasonically vibrated hearth called a "sonotrode" that is attached to a booster and transducer with a piezo electric element.[5] An 80–400 A arc melts the material placed on the sonotrode. The 20 kHz vibrations through the sonotrode create standing capillary waves on the surface of the liquid metal.[6] When the amplitude of the wave breaks the intermolecular forces of the liquid, droplets of molten metal are ejected from the pool and assume a spherical shapes because of surface tension. The droplets are carried away in the stream of low-pressure argon and solidify before touching any surface, thereby producing high purity spherical powder. The size of droplet ejected from standing capillary waves can be approximated using Lang's equation:[7]

$$D_p = 0{,}34\left(\frac{8\pi\sigma}{\rho f^2}\right)^{1/3} \quad (1)$$

where D_p is the droplet size [µm], σ is the surface tension [N/m], ρ is the density of liquid [kg/m3], and f is the frequency [kHz].

The frequency of ultrasonic vibrations will shift the particle size distribution produced. Higher frequency produces finer droplets with other potential applications.

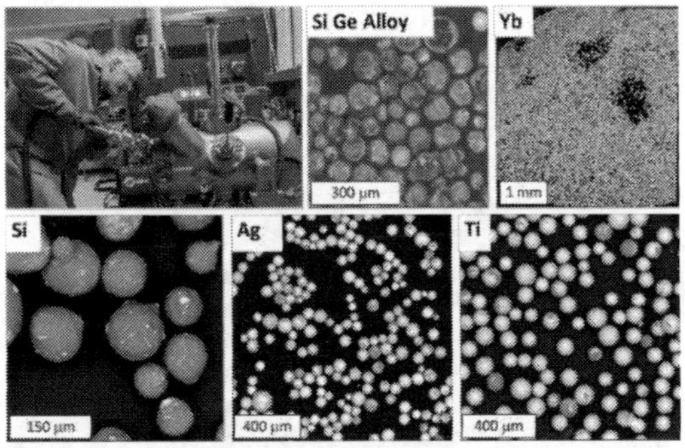

Fig. 7. Various metals produced by ultrasonic atomization. Upper left shows the instrument in operation.

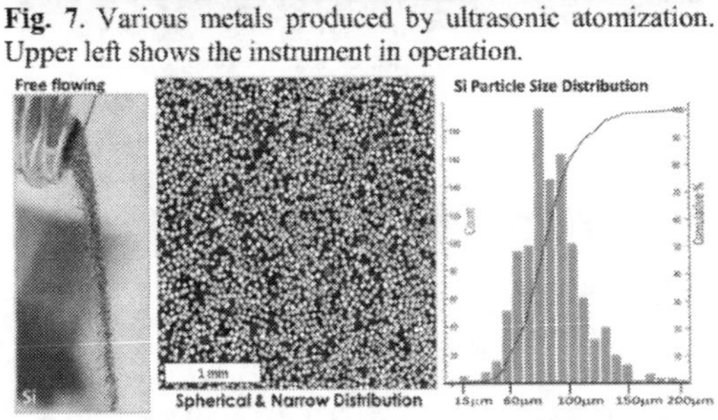

Fig. 8. Early production of powders from the AMAZEMET rePowder ultrasonic atomization tool. (Left: Solid silicon flowing like a liquid from a vial; middle: silicon powder produced; right: size distribution of particles n = 1049).

Analysis of each powder shows a high degree of uniformity. Fig. 8 illustrates the properties of being free flowing, spherical, and uniform in size, which are important for many applications. The large particles and unatomized material were removed using a sieve with a 200 µm mesh opening. Metal vapor, which occurs with arc melting, condenses to coat the inside of the chamber. Typically, this submicron, light, fluffy material is removed by shaking, sonication, and rinsing before drying. Tiny, high surface area particles are removed, which improves the chemical properties.

4 Powder applications

Spherical powders have relatively unexplored specific applications compared to traditional powders:

- Microgram precision dispensing—Fig. 9a shows 19 tungsten particles totaling 21 µg. Eventually, single particle dispensing may be possible. Spherical powders can be easily dispensed without loss from particles clinging to the vessel.
- Improved fabrications from consistent material and chemical properties.
- Making targets with open architecture allowing rapid dissolution and new cooling options—Fig. 9b and 9c show a platinum ruthenium pellet that has been sintered the tangents of the spheres. A solution has nearly instant access to every particle.
- Additive manufacturing and some nuclear targetry applications. Fig. 9d shows Inconel 625 alloy being fused by a laser. This is enabled by a uniform height of powder being drawn with each subsequent layer.

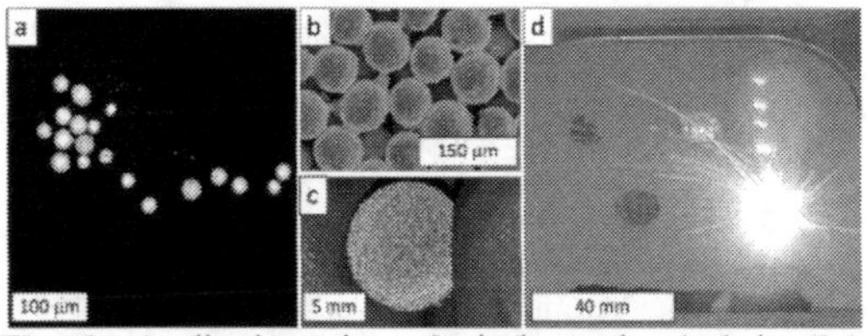

Fig. 9. Applications for spherical powder include **(9a)** precision micro dispensing, **(9b)** fabrication of unique sintered materials with open architecture, **(9c)** high flow rates and quicker dissolution, and **(9d)** applications for solid metals that can flow like a liquid including making thin layers of powder beds for additive manufacturing applications.

This work is supported by the US Department of Energy Isotope Program, managed by the Office of Science for Isotope R&D and Production.

References

1. J. Grubbs, et.al., Metals **12** (4), 603 (2022).
2. E. A. Seddon, Chemistry of Ruthenium. (1984).
3. J. Knauer, et. al. Oak Ridge National Laboratory, (1993).
4. R. A. Dunlap, et.al., Canadian Journal of Physics **78** (3), 211 (2000).
5. Ł. Żrodowski, et. al., Materials **14** (10), 2541 (2021).
6. E. G. Lierke and Griessha.G, Ultrasonics **5** (OCT), 224 (1967).
7. R. J. Lang, Journal of the Acoustical Society of America **34** (1), 6 (1962).

https://doi.org/10.1051/epjconf/202328503003

Development of low-loss minimal exposure technique for thallium foil fabrication

Jennifer Conner[1],* *Mike* Zach[1], and *Donald* Blackburn[2]

[1]Stable Isotope Materials and Chemistry Group, Oak Ridge National Laboratory, Oak Ridge, Tennessee, USA
[2]Health and Safety Programs Group, Oak Ridge National Laboratory, Oak Ridge, Tennessee, USA

Abstract. Despite its availability as a solid bead or ingot, non-isotopic thallium metal is no longer commercially produced as a foil because of its high toxicity. To conduct fundamental studies on the material, the Stable Isotope Materials and Chemistry Group (SIMC) at Oak Ridge National Laboratory (ORNL) was approached to develop a safe method to process thallium and produce 40 non-isotopic thallium foils. The commercially sourced metal was consolidated into an ingot by melting the material in a special tube furnace under a reducing atmosphere. The resulting ingot was cold rolled using a work-hardened, stainless-steel pack and oil lubricant before cutting the final foils to 2.5 × 2.5 cm^2 area and thickness of 50–75 mg/cm^2. The appropriate safeguards used at each step are outlined to ensure the safe and consistent production of high-quality foils. The low-loss process enables future requests for isotopic thallium and other hazardous and rare materials.

1 Introduction

In June 2021, The Stable Isotope Materials and Chemistry (SIMC) group at Oak Ridge National Laboratory (ORNL) was requested to develop a method to fabricate 40 natural thallium metal foils, each sized at 2.5 x 2.5cm^2 area and thickness of 50 – 75 mg/cm^2, for applications in the production of radioisotopes at national laboratories across the United States. SIMC is well-established as one of the major stewards and producers of stable, isotopically enriched targets in the United States. Because of the highly limited supply and high cost of the materials used, many of the projects executed require extremely low-loss and high-efficiency processes. Herein, the specific difficulties associated with thallium metal are explored, especially as they relate to health, safety, and contamination concerns. The fabrication procedure used for the completion of this specialized task is also outlined.

2 Health and safety concerns

Thallium metal is known for its high acute toxicity and how easily it absorbs through the skin and contaminates other metals and materials [1]. Sigma Aldrich sets its workplace control parameters at a 0.1 mg/m^3 time-weighted average (TWA), half of the parameter for lead, and a dermal exposure of 0.02 mg/m^3 TWA [2]. Exposure to skin and mucous membranes results in an 80–100% absorption of the metal [3-4]. In adults, the lethal dose has been found to vary between 6 and 40 mg/kg; however, acute effects begin well below this range.

The excretion of thallium from the body has an estimated half-life of approximately 10–30 days, which increases the danger of continued exposure to the material. The high acute toxicity of this metal is why commercial suppliers have largely forgone processing thallium on a large scale. Thus, the process development of this project was partially dedicated to finding ways of working safely with thallium.

2.1 Work controls

According to the safety data sheet [2], the suggested engineering controls include a face shield and safety glasses, wearing nitrile gloves along with vigorous handwashing, a complete Tyvek or Tychem suit, and either a full-face air supplying respirator or a full-face particle respirator, as determined by risk assessment. In this procedure, individuals must dress out fully, donning double nitrile gloves (taped at the wrists), chemical resistant suits, and powered air purifying respirators (PAPRs), and they must work within a chemical hood as often as possible and limit access to the room. All work is conducted through a buddy system to account for human error.

2.2 Health and safety monitoring

Individuals working with thallium were placed on a thallium monitoring program through the Health and Safety Program at ORNL. When individuals are actively working with the material, biological sampling at the end of the work week is imperative to account for cumulative dosing during normal work conditions. Because cumulative dosing may also result in chronic intoxication, the individuals were

* Jennifer Conner: 9dc@ornl.gov
Notice: This manuscript has been authored by UT-Battelle, LLC, under contract DE-AC05-00OR22725 with the US Department of Energy (DOE). The US government retains and the publisher, by accepting the article for publication, acknowledges that the US government retains a nonexclusive, paid-up, irrevocable, worldwide license to publish or reproduce the published form of this manuscript, or allow others to do so, for US government purposes. DOE will provide public access to these results of federally sponsored research in accordance with the DOE Public Access Plan (http://energy.gov/downloads/doe-public-access-plan).

monitored for at least 90 days after the conclusion of the project [5].

Airborne particulate was monitored by NIOSH 7303 sampling. NIOSH 7303 has a limit of detection of approximately 0.044 µg/mL in a 25 mL sample. The amount found in the four samples was found to be below the limit of detection, thus the PAPRs will not be necessary in future iterations of the process if all steps are followed as described herein.

Surfaces that could be contaminated during processing were sampled using the IOP 01-12.05 [6] method and analyzed using EPA Method 6010C [7] to determine if thallium had contaminated the lab space. IOP 01-12.05 is a method traditionally used for toxic metal contamination collection by surface wipe sampling. Because of the nature of the sampling method, it is used as a test for removable contamination. It was decided that individuals working in spaces with unconfirmed levels of fixed contaminate would continue to be monitored yearly.

3 Fabrication of thallium foils

3.1 Specialized equipment

Two pieces of equipment were developed: a tube furnace tube with a water-cooled copper mesh and coil vapor capture system to combat the high vapor pressure of molten thallium (Fig. 1) and an argon shower box to prevent oxidation of the reactive metal (Fig. 2). As a safety precaution, both items were assembled to fit inside the available chemical hood during processing and to be disposed of after processing.

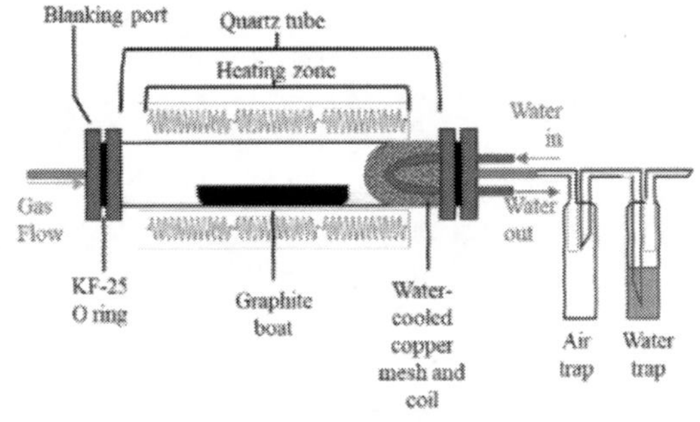

Fig. 1. Schematic of disposable tube furnace.

Fig. 2. Argon shower box with glove inserts.

In addition to being sized to fit the chemical hood, the tube furnace included a water-cooled copper mesh to act as a thallium vapor collector. In line with the collector, a water trap was used at the outlet of the tube. These traps prevent backflow of any oxidizing species in air.

3.2 Melting

The natural material was received from the supplier as 100 g of 6 mm diameter shot. The beads arrived under an aqueous salt solution. After rinsing them with ethanol and wiping clean, they were transferred to pre-dried graphite boats (Fig. 3A) and introduced to the tube furnace. After being sealed and leak checked, the volume of the quartz tube was purged with argon for 4× the volume of the tube to ensure a fully inert atmosphere. To reduce any oxide that may have formed on the thallium shot during transfer, hydrogen was introduced to the tube during the melt.

Using the tube furnace, the material was rapidly heated to 320°C and consolidated to an ingot (Fig. 3B) of ~5 cm length. Minimal amounts of thallium were lost to vaporization caused by the rapidity of the melting, and the resulting ingot appeared shiny and oxide-free. Three large ingots ~10-12 cm long were made using the same process to improve final storage of the material and use in rolling the material to a foil.

Fig. 3. Thallium in the tube furnace [A] as shot, before melting, and [B] as an ingot.

3.3 Cold rolling

The resulting ingots oxidized quickly when removed from the inert atmosphere of the tube. One ingot quickly turned dull and black. The material was always kept in the argon box unless it was actively in process. In accordance with the "As Low As Reasonably Achievable" principles, it was decided that the reactive metal would be rolled in open air in the largest available rolling mill. This allowed for the maximum amount of material to be rolled per pass, decreasing prolonged exposure to the metal by orders of magnitude and made cleanup and post-fabrication sampling more accessible. The metal was rolled inside a work hardened stainless steel rolling pack. Because this process could not be done under inert atmosphere or vacuum due to the size and accessibility of the chosen rolling mill, the thallium metal was covered in an oil to act as a lubricant in the pack, to slow oxidation of the surface, and to capture any of the flaky, black oxide that could escape as a toxic particulate.

The metal was rolled out to 90 mm in width before being turned 90° and rolled on the perpendicular axis until the piece approached the customer's desired thickness (Fig. 4). Thallium proved to be gummy and soft, with a tendency to adhere to the pack. The lubricating oil was key to keeping the metal mobile, and periodic shifting of the foil and reapplication of the oil within the pack were essential. A lower force per unit area supplied by the rolls also assisted in preventing adherence to the pack. Like many other soft metals, thallium was expected to stick to the pack more readily when passing through smaller diameter rolls.

Fig. 4. Thallium foil, rolled in a stainless-steel rolling pack.

Despite these precautions, the metal was still fragile and tore easily during handling and processing. During rolling, if the rolling pack was passed through the mill seam first, then the curved pack could spring open when it was released from the rolling mill rolls and tear the foil. Instead, the pack was rolled so the seam would be last to pass through the active area of the mill, minimizing the chance of snapping the foil. Although this is not an option for many metals, the malleability of thallium allowed it to be easily rolled without deforming the pack.

3.4 Cutting and sealing

Once the correct areal density was assured, the material was meticulously cut to shape (Fig. 5) using a 2.50×2.50 cm^2 template, cleaned of oil using a Kimwipe and ethanol, and had its dimensions recorded while within the argon shower box.

To seal the foils, a piece of 2 in. wide Kapton tape was placed adhesive side up. The foil was placed with one edge in contact with the tape, and allowed to fall toward the center, where it was carefully pressed into the tape. A second piece of tape was bent in a U shape, brought into contact with the thallium foil, and gently pressed into the bottom piece of tape. Air bubbles were removed using gentle pressure and a straight edge. The tape was then cut to a 2.50×2.50 cm^2 square and wiped clean with an ethanol-wetted Kimwipe.

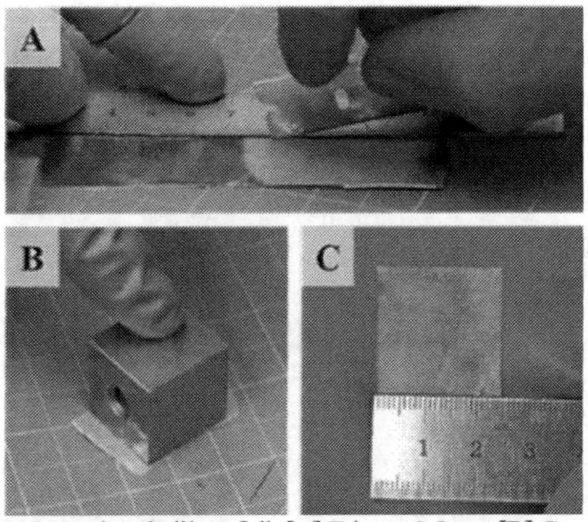

Fig. 5. Cutting thallium foil. [A] Trim to 2.5 cm. [B] Cut to 2.50×2.50 cm^2 using template. [C] Final thallium foil.

After sealing between Kapton tape, each sealed foil was transferred to a vacuum chamber to remove any remaining air sealed within the Kapton tape before mounting. The target was placed carefully on the frame—with the foil at the center—and the corners of the tape were secured by 0.25 in. sized Kapton dots. Fig. 6 shows the final products.

Fig. 6. Forty-two thallium foils mounted to frames and packaged to be sent to customers.

3.5 Mitigation of surface contamination

To reduce the amount of thallium remaining on laboratory surfaces and to return equipment to a usable state, all surfaces were sampled using surface wipes per IOP 01-12.05. The surface wipes were analyzed via the EPA Method 6010C, which has an estimated limit of detection at approximately 27 µg/L, within ±20% of the true value. Several surfaces within the possible contamination area were tested using this method (Table 1). According to the OSHA Technical Manual, *Surface Contaminants, Skin Exposure,*

Table 1. IOP 01-12.05 and EPA 6010C removable surface contamination results.

Sample location	Contamination (µg/cm^2)
Hood sash	<10
Working surface in hood	27
Right side wall in hood	<10
Small glove box in hood (outside)	14
Small glove box in hood (inside)	<10
Lab floor (by hood)	<10
Rolling mill (top roll)	180
Rolling mill (bottom roll)	400
Chair used during fabrication	<10
Field blank	<10

Monitoring and Other Analyses, the rule of thumb is to use the maximum allowable dose from skin contact (20 μg/cm^2) over 100 cm^2 (for a total of 200 μg/100 cm^2) as a pass/fail for the contamination of the tested surface. Table 1 reveals that the rolling mill rolls and the chemical hood exceeded this value and thus prompted additional decontamination efforts.

Additional cleaning measures were implemented to ensure the rollers of the Stanat 2-HI mill and the chemical hood were decontaminated to an appropriate level for continued work. The rollers were wiped repeatedly with a shop towel soaked in Formula 409 solution. The surface of the rollers was then wiped with Scrubs Stainless Steel Cleaning Wipes until the rollers returned to their pre-work condition. The conditioner in the wipes was allowed to soak into the rolls overnight. Afterward, ethanol and a shop towel were used to repeatedly wipe the rollers until the towel came away clean.

The amount of thallium contamination on the rollers was decreased significantly after conducting the rigorous cleaning protocol (Table 2). None of these limits reached the set limit of 200 μg/100 cm^2. Because of the presence of thallium remaining on the rollers, the Stanat 2-HI mill was designated as potentially contaminated until a definitive test could verify no fixed contamination was present, or the rollers are replaced.

Table 2. IOP 01-12.05 and EPA 6010C removable contamination results after cleanup.

Sample location	Contamination (μg/cm^2)
Working surface in hood	<10
Rolling mill (top roller)	19
Rolling mill (bottom roller)	19
Field blank	<10

4 Summary

SIMC at ORNL was able to develop a process for ensuring the fabrication of high quality, high demand natural thallium metal foils from commercially available material for use in national labs across the United States as well as ensuring the safety of the personnel working with this toxic element. All 42 foils produced were within customer-supplied specifications, affixed to customer supplied frames for use in their experiments, and successfully shipped to the customer. No individuals monitored during or after the completion of this campaign showed biological thallium concentrations above background and all surfaces showed near-background concentrations of removable thallium contamination.

References

1. National Institute for Occupational Safety and Health, Thallium: Systemic Agent, https://www.cdc.gov/niosh/ershdb/emergencyresponsecard_29750026.html (2003).
2. Thallium Safety Data Sheet, Version 6.6, Sigma-Aldrich (2021).
3. M.S. Sinicropi., A. Caruso, A. Capasso, C. Palladino, A. Panno, C. Saturnino, Heavy metals: Toxicity and carcinogenicity. Pharmacologyonline 2010, 2, 329–333, (2010).
4. G. Genchi, A. Carocci, G. Lauria, M.A. Sinicropi, A. Catalano, Thallium Use, Toxicity, and Detoxification Therapy: An Overview, Appl. Sci. 11, 18, 8322 (2021).
5. Health Monitoring for Thallium, https://www.safeworkaustralia.gov.au/resources-and-publications/guidance-materials/health-monitoring-thallium (2020).
6. L.G. Greely, J. Sicau, Health and Safety Group Sampling Program Procedure: Surface Sampling for Metals, IOP 01-12.05, Rev. 2 (2017).
7. Method 6010C, https://archive.epa.gov/epawaste/hazard/testmethods/web/pdf/method%206010c%2C%20revision%203%20-%202007.pdf (2007).

UO_2 target preparation with spin coating assisted solution combustion synthesis

Ashabari Majumdar[1*], *Khachatur* V. Manukyan[1], *Stefania* Dede[1,2], *Jordan* M. Roach[3], *Peter* C. Burns[4], and *Ani* Aprahamian[1,5]

[1]Nuclear Science Laboratory, Department of Physics, University of Notre Dame, Notre Dame, IN 46556, USA
[2]Cyclotron Institute, Texas A&M University, College Station, TX 77843, USA
[3]Department of Chemistry and Biochemistry, University of Notre Dame, Notre Dame, IN 46556, USA
[4]Department of Civil and Environmental Engineering and Earth Sciences, University of Notre Dame, Notre Dame, IN 46556, USA
[5]A. Alikhanyan National Science Laboratory of Armenia, 2 Alikhanyan Brothers, 0036 Yerevan, Armenia

Abstract. A simple and efficient target preparation method is developed combining spin coating and solution combustion synthesis. Multiple smooth and uniform UO_2 targets have been prepared using this method on a variety of backings (aluminium, carbon, silicon) used in nuclear physics experiments. The thicknesses of the targets can be precisely tuned by changing the number of coatings within the range of ~50-1000 μm/cm^2. These targets are highly uniform (<5% deviation), robust, and remain strongly adherent to their backings even after being irradiated by high doses (10^{17} ions/ cm^2) of 1.7 MeV Ar^{2+} ions.

1 Introduction

High quality, robust, and uniform actinide targets are important for a broad range of studies from nuclear astrophysics to the stockpile stewardship.
For obtaining accurate experimental results, uniform, crack- and contamination-free actinide targets are required. These should also be stable, structurally robust and adhere to the target backings and withstand high beam irradiation during experiments.

The various methods available for actinide target preparation have certain advantages and disadvantages. For example, mechanical processes such as cold rolling of metallic actinides pose safety hazards due to the pyrophoric nature of the metals [1]. In physical methods, vacuum evaporation requires a complex and high vacuum experimental set-up. The efficiency for this method is low and not suitable for making targets with expensive actinide materials [2]. The targets prepared by chemical methods (electrostatic deposition and molecular plating) often exhibit low stability under irradiation due to poor electrical and thermal conductivity [3]. Another proposed method for actinide target preparation is polymer assisted deposition (PAD). Although, PAD produces targets of high uniformity, this process requires long annealing periods at high temperature (1000 ºC) post deposition and specific backing materials [4].

At the University of Notre Dame, we have developed a simple and cost-efficient method for uranium dioxide (UO_2) target preparation using spin coating assisted solution combustion synthesis. The UO_2 targets are produced on different backings, and are found to be uniform, smooth and homogeneous. The target thicknesses can be well tuned within the ~50-1000 μm/cm^2 range. These show high thickness uniformity (<5% deviation) within a spot diameter of 25 mm. The targets are also found to not disintegrate or lose material under high intensity irradiation. The material yield of this method ~30%. In our previous work, we discussed the details of irradiation induced structural changes in UO_2 and $UO_{2.12}$ targets on aluminium backings [5-8]. In the current work, we provide a summary of this novel target making method for making highly uniform UO_2 targets that can be extended for preparing other actinide targets with a wide range of thicknesses on a variety of backings.

2 Experimental Methods

The optimal combination of the precursors (reactive solution, fuel and solvent solution) for the UO_2 target fabrication using solution combustion synthesis is determined using differential scanning calorimetry (DSC) coupled with thermogravimetric analysis (TGA) [7]. In this case, the reactive solution, fuel and solvent are isotopically depleted uranyl nitrate hexahydrate ($UO_2(NO_3)_2 \cdot 6H_2O$, International Bio-Analytical Industries, ≥98.0%), acetylacetone fuel ($C_5H_8O_2$, TCI, >99.0%) and 2-methoxyethanol ($C_3H_8O_2$, Alfa Aesar, 99%) respectively. The final working solution is a 0.25 M uranyl nitrate solution in a 0.5 molar ratio with the fuel.

A volume of 100 μL of the uranyl solution is spin coated on the atmospheric plasma treated backings at a

* Corresponding author: amajumda@nd.edu

speed of 3500 rpm/s for 35 seconds with a spin coater (SPIN 150i NNP). The angular acceleration is 100 rpm/s. Then the samples are annealed for 20 minutes at 400 °C in a furnace, in order to produce the UO_2 layers. With this procedure, targets are produced on a variety of backings (aluminium alloy 6061, 99.997% pure aluminium, carbon and silicon) that are widely used as backings in nuclear experiments. The dimensions of both McMaster-Carr 6061 aluminium alloy and 99.997% puratronic grade Alfa Aesar aluminium discs are 42 mm in diameter and 0.7 mm in thickness. Carbon foils from ACF Metals used as backing are made of 98.9% ^{12}C and 1.1% ^{13}C. The dimensions of carbon backings made from these foils are 25 mm × 25 mm × 0.3 µm. The p type (boron doped) silicon wafer backings have a diameter of 50.8 mm and thickness of 0.3 mm. To prepare targets with different thicknesses, the spin coatings and combustion are repeated 1, 3, 5 and 7 times on aluminium alloy backings. The target made on pure aluminium is coated with the solution followed by a combustion of 30 times. In addition, three targets with identical parameters are made on aluminium alloy to investigate the reproducibility of the procedure.

The surface morphology of the targets is measured with a dual electron/ion beam Helios Nanolab 600 (FEI – Thermo Fisher Scientific). Their cross-sectional images are captured after depositing a protective platinum layer on top of the target layers using Helios. An Orbis (EDAX) X-ray fluorescence (XRF) spectrometer equipped with a Rh X-ray tube, poly-capillary optics, a sample stage with automated XYZ positioner and an 80 mm^2 Si(Li) drift detector is used to map the compositions of the targets and determine the uniformity as well as elemental abundances of the targets. The mapping is done by attaching the targets to the stage and scanning the target surface with a 30 µm X-ray beam spot under vacuum (~6 Pa absolute). The spatial resolution of images obtained from the mapping is ~160 µm. The experimental conditions are 40 keV and 400 mA with 0.6 second sampling time per spot. For analyzing elemental abundances, spot scans are done over the area of the targets with a 2 mm beam spot and 100 live seconds for each spot. An α-particle spectrometer (ORTEC-ULTRA-AS) with an ion-implanted silicon detector (active area – 900 mm^2, resolution - 29 keV at 5.486 MeV) is used to determine the absolute amounts of uranium isotopes in the targets prepared. The details of the calibration of the instrument and data analysis method are described in our previous work [7].

3 Results

The essential characteristics for targets are to be uniform and smooth, which are necessary to produce reliable data during nuclear experiments. The targets on aluminium, carbon and silicon backings prepared with spin coating assisted combustion method are found to be smooth without any defects. The nanoscale images of the cross sections of these target layers on all three different backings are uniform and pore-free (figure 1 A-C).

Fig. 1. Nanoscale images of the cross sections of uranium dioxide layers obtained on aluminium (A), carbon (B) and silicon (C) backings using the spin coating assisted solution combustion synthesis method. The images show smooth and uniform cross sections. Platinum is used for imaging purposes.

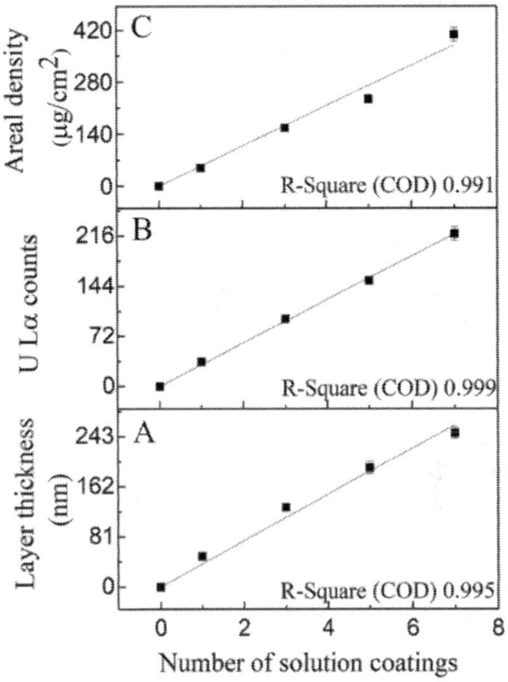

Fig 2. Number of solution coatings in function of layer thickness (A), U Lα counts (B) and areal density (C) derived from SEM, XRF and α–particle spectroscopy measurements, respectively. It is shown that the uranium content in the targets increases linearly with increasing number of coatings of the solutions.

The thicknesses of the UO_2 target layers are measured from the nanoscale cross section images. The target on the puratronic grade aluminium backing with 30 coatings is the thickest (~1000 µm/cm²). The thicknesses of the 1, 3, 5 and 7 layered targets on aluminium alloy increase linearly with the increased number of layers (figure 2A). Counts of characteristic uranium X-ray lines (U Lα) are measured using XRF and are plotted (figure 2B) with respect to the number of target layers. The linear trend among them resembled the previous one in figure 2A. The U Lα counts are obtained from averaging 2 mm beam spot scans of 17 different points of each target, and the standard deviation in each case is less than 5%. A third method, α-particle spectroscopy is used to confirm a linear trend on an even larger (~cm) scale. Alpha-particles emitted from the targets are measured and used to determine the areal density of the total uranium layer in µg/cm^2. The areal density of the total uranium layer also showed a similar trend with respect to the number of layers as seen from the SEM and XRF techniques (figure 2C).

We also investigated the overall uniformity of these targets using an XRF mapping. As seen in the first column of the figure 3, the blank target does not show any uranium content. Absence of bright green spots (U M lines) anywhere on the backing or the XRF stage indicates that there is no detectable uranium. Al K lines emitted from the aluminium backing are denoted by red. In the targets with 1, 3, 5 and 7 coatings, the bright green circular spot at the centre of the backing signifies the presence of uranium. The brightness of the green spots increases with the number of coatings. At the same time, the intensity of the Al K line distribution decreases with increasing coatings. Lesser number of characteristic X-rays of aluminium can be detected in these cases due to the higher content of uranium on the aluminium backing.

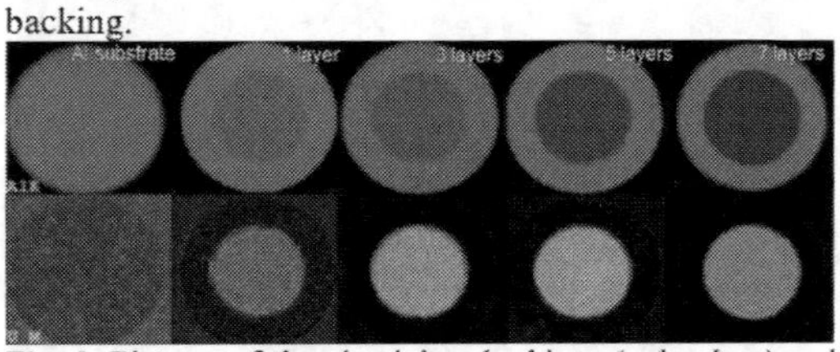

Fig. 3. Pictures of the aluminium backings (red colour) and uranium layers (green colour) derived by XRF mapping showing an overall uniformity of the UO_2 targets with different layers.

Three UO_2 targets prepared with identical parameters are also mapped using the same method and they show a uniform presence of U M lines in the active target region. The intensity of the U M lines is the same for each, which suggests similar thicknesses. A same amount of decrease in the intensity of the Al K lines in the target region confirms that as well. Several spot scans are done all over these three targets and the standard deviation of the U Lα counts among them is less than 1%.

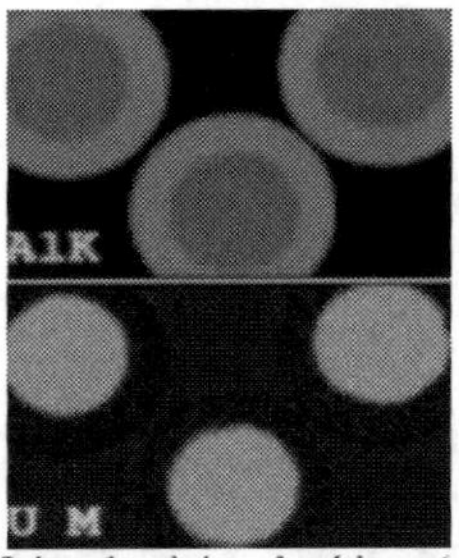

Fig. 4: Pictures of the aluminium backings (red colour) and uranium layers (green colour) of three UO_2 targets prepared with identical parameters derived by XRF mapping showing a reproducibility of the target preparation method.

These targets on aluminum backings are irradiated with 1.7 MeV energy Argon ions (Ar^{2+}) to a fluence (~10^{17} ion/cm^2) under a controlled temperature. The uranium content on the targets is checked using α-particle spectroscopy after irradiation and no significant material loss is found.

4 Discussion and Conclusion

Several UO_2 targets are produced using a spin coating assisted combustion technique on different backings (aluminium, carbon and silicon) used in nuclear experiments. In each case, the surfaces as well as the cross sections are found to be smooth and uniform. A linear relationship is determined between the number of layers coated on the backings and the final UO_2 content on the targets using SEM, XRF and α-particle spectroscopy. This target preparation method provides a precise control of the target thicknesses within the ~50-1000 μm/cm² range. The targets produced are highly reproducible. The targets are exposed to high doses of Argon ion irradiation (~10^{17} ion/cm^2). They did not disintegrate or lose adherence to the backing during the experiment. The total uranium content of the targets is also found to remain the same after the irradiation. Overall, this novel method proves to be an excellent candidate for making UO_2 targets of a variety of thicknesses on different backings and can potentially be extended for other actinide target making.

The work is performed with financial support from the U.S. Department of Energy's (DOE) National Nuclear Security Administration (NNSA, Grant # DE-NA0003888 and DE-NA0004093), National Science Foundation (NSF, PHY-1713857), and Nuclear Science Laboratory (Grant # PHY-2011890), Notre Dame, JINA-CEE NSF Physics Frontiers Centers (Award #: PHY-1430152), US. The authors also acknowledge the Center for Environmental Science and Technology (CEST), Materials Characterization Facility (MCF), Notre Dame Integrated Imaging Facility (NDIIF) for instrumental usage.

References

1. H. J. Maier, Nucl. Instrum. Methods Phys. Res. **200**, 113 (1982)
2. A. Stolarz, R. Eykens, A. Moens, Y. Aregbe, NIM A **613**, 351 (2010)
3. I. Usoltsev, R. Eichler, R. Dressler, D. Piguet, D. Wittwer, A. Turler, R. Brursch, E. A. Olsen, J. P. Omtvedt, A. Semchenkov, NIM A **691**, 5 (2012)
4. M. A. Garcia, M.N. Ali, N. N. Chang, T. Parsons-moss, P. D. Ashby, J. M. Gates, L. Stavsetra, K. E. Gregorich, H. Nitsche, NIM A **613**, 396 (2010)
5. S. Dede, K. V. Manukyan, J. M. Roach, A. Majumdar, P. C. Burns, A. Aprahamian, Appl. Surf. Sci. **603**, 154437 (2022)
6. J. M. Roach, K. V. Manukyan, A. Majumdar, S. Dede, A. G. Oliver, P. C. Burns, A. Aprahamian, Inorg. Chem. 60, 18938 (2021)
7. A. Majumdar, K. V. Manukyan, S. Dede, J. M. Roach, D. Robertson, P. C. Burns, A. Aprahamian, ACS Appl. Mater Interfaces 13, 35153-35164 (2021)
8. A. Majumdar, K. V. Manukyan, W. Tan, S. Dede, J. M. Roach, A. Couture, P. C. Burns, A. Aprahamian, NIM A 1045, 167551 (2023)

Actinide and lanthanide thin-layer developments using a drop-on-demand printing system

Dennis Renisch[1,2*], *Christoph E.* Düllmann[1,2,3], *Roman* Abrosimov[1], *Klaus* Eberhardt[1], *Birgit* Kindler[3], *Bettina* Lommel[3], *Jana* Lukacova[1], *Christoph* Mokry[1,2], *Stephan* Oberstedt[4], *Ilkka* Pohjalainen[5], *Jörg* Runke[1,3], *Tanyel* Schwab[1], and *Christopher* Sirleaf[1]

[1]Johannes Gutenberg-Universität Mainz, 55099 Mainz, Germany
[2]Helmholtz-Institut Mainz, 55099 Mainz, Germany
[3]GSI Helmholtzzentrum für Schwerionenforschung GmbH, 64291 Darmstadt, Germany
[4]European Commission, Joint Research Centre, 2440 Geel, Belgium
[5]University of Jyväskylä, 40014 Jyväskylä, Finland

Abstract. Actinide and lanthanide thin layers with specific requirements regarding thickness, homogeneity, chemical purity, mechanical stability, and backing properties are applied in a multitude of physics and chemistry experiments. A novel target preparation method, the so-called "Drop-on-Demand" (DoD) technique, based on a commercial nanoliter (nL) dispenser is applied since a few years in the Nuclear Chemistry unit at Johannes Gutenberg University Mainz. The wetting behaviour of the nL droplets on the substrate's surface is a key parameter determining the spatial distribution of the deposited material after evaporation. By switching from aqueous to organic solvents as well as by substrate surface modifications, the wetting behaviour can be influenced. Recent investigations on this influence and applications of the DoD method are presented. The produced actinide deposits were characterized by optical and scanning electron microscopy, by α spectroscopy as well as by radiographic imaging.

1 Introduction

Radioactive targets and sources are needed for a variety of physics experiments. The applications range from targets for laser [1] and mass spectroscopic investigations of actinides [2], over myon-based experiments for the determination of nuclear charge radii [3] to recoil sources where the daughter nuclides are to be used for atomic physics [4], nuclear physics [5] and chemical [6] experiments, to targets used for superheavy element production in heavy ion induced fusion reactions [7].

The requirements of the target properties are as diverse as the applications. In addition to the desired nuclide and quantity, the layer thickness, homogeneity, geometry, chemical form, mechanical stability and the substrate used are the most important parameters. All these parameters have to be tailored to the requirements of the individual experiment, so that no single, general specification for "the optimum target" can be given. Accordingly, the fabrication methods must also be flexibly adapted. For many applications, electrochemical deposition, such as molecular plating [8], is a viable method. Disadvantages of molecular plating include the need to use a conductive substrate, the difficulty of obtaining a chemically well-defined layer and limitations in the applicable geometry.

At the Johannes Gutenberg University Mainz, the Drop-on-Demand (DoD) technique has been used for several years as a complementary target fabrication method [9]. The DoD system essentially consists of a commercial nanoliter (nL) dispenser (PipeJet® Nanodispenser from BioFluidix), which is mounted above movable x-y stages, jointly controlled by a versatile control system.

The substrate is positioned on the movable stages. A self-written LabView-based software allows printing any geometries. The printing solution, in which the radioisotope of interest is dissolved, is filled into a reservoir at the dispenser. Aqueous as well as organic solutions can be used. By combining different solvents and different backings (metals, polymers, carbon substrates, etc.), these surfaces can be modified [10,11] and different wetting behavior of the individual drops can be achieved. This also has to be adapted and optimized for the respective target, depending on whether individual, separated deposits are required or whether the backing needs to be covered over as large an area as possible.

In the following, the further development of the method is described on the basis of selected examples from recent years, in which DoD produced targets were used, in particular with regard to the use of different solvents for printing.

* Corresponding author: renisch@uni-mainz.de

2 Experimental results

2.1 Comparison of aqueous and organic solvents

2.1.1 Spontaneous fission sources

For a project in collaboration with JRC Geel, Belgium, to determine fission fragment masses and kinetic energy distributions by means of the double velocity and double energy (2v-2E) method, ^{248}Cm and ^{252}Cf sources are needed. The fission products will be measured in 4π geometry using the VERDI spectrometer [12]. For high-precision data, the energy loss of the fission products inside the source layer and substrate has to be as small as possible. For this experiment, a rate of about 100 to 1000 fissions/s of ^{248}Cm or ^{252}Cf was requested. The source thus has to be as thin as possible, and a particularly thin substrate is required. The substrate of choice is a 35 µg/cm² polyimide (PI) foil, covered with a 50 µg/cm² gold coating. Due to the fragility of the foils, mounting into an electrochemical cell is not possible due to the risk of rupture during mounting or dismounting, thus the contactless DoD process was used to fabricate the sources.

For the first attempt to produce a suitable source, ^{252}Cf was printed from an aqueous solution. This involved printing of 221 drops, 20 nL each, 1 mm apart, onto a PI substrate. The source contained a total of 330 kBq α activity and 10 kBq spontaneous fission (sf) activity. The relatively high hydrophobicity of the substrate led to single, spherical droplets, resulting in comparatively thick deposits after drying. The sources were tested at JRC Geel, where it was found that the energy loss within the source material was significantly greater than the energy loss in the substrate. Due to the resulting poor resolution, the sources were not suitable for the desired measurements.

To overcome these limitations, different alcohol-based solvents were tested, in particular ethanol, isopropanol and isobutanol. Preparatory experiments using lanthanide material spiked with ^{243}Am to allow for studying deposit thickness by measure of α particle energy loss showed that an ethanol:water = 72:25 mixture appears to be well suited [13]. In this case, the droplets cover a much larger area than when printing from pure water, but the area to be wetted remains controllable. At higher alcohol content, no defined wetting of the substrate could be obtained. By distributing an identical amount of source material over a larger area, the average film thickness should be significantly smaller than in the previous sources.

^{248}Cm sources were prepared using a printing sequence of 641 drops, 10 nL each, using an ethanol:water = 70:30 mixture as solvent. Fig. 1 shows a microscope image (left) and a radiographic image (right) of one of the finished ^{248}Cm sources. The source contains 8.1(3) µg of ^{248}Cm. An α activity of 1270(38) Bq was measured, corresponding to a spontaneous fission rate of 114(4) Bq. Alpha spectroscopic measurements show a relatively narrow FWHM of 19 keV, which is close to the intrinsic resolution of the used PIPS detectors [14].

The characterization of the fission fragment energy distributions is ongoing at JRC Geel.

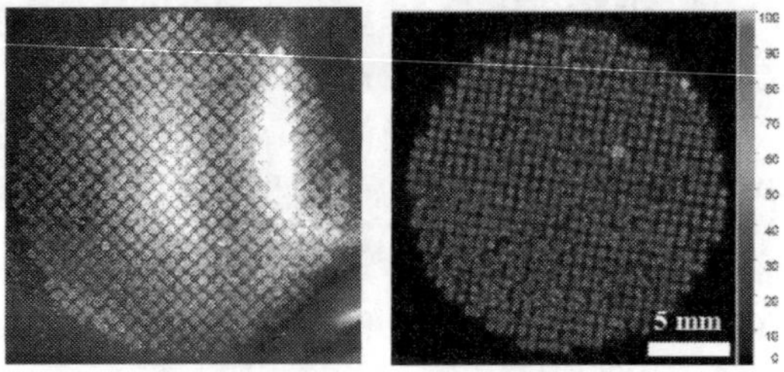

Fig. 1. Microscope image (left) and radiographic image (right) of a 8 µg ^{248}Cm source, printed with the DoD setup from a mixture of ethanol:water = 70:30. The source was produced by placing 641 drops of 10 nL each on a Au coated PI substrate [14].

2.1.2 Targets for accelerator-based experiments

At the university of Jyväskylä, the properties of n-deficient Th to Pu isotopes shall be studied at IGISOL [15]. Those isotopes are produced by p-induced reactions on neutron-poor actinide targets, like ^{232}Th or ^{233}U. Thus, nuclides in the range $^{226\text{-}229}$Th, ^{227}Pa, or $^{224\text{-}233}$Np are accessible. Many fundamental properties of the nuclides, that can be reached this way, have hardly been investigated so far or are based only on extrapolated values. For example, direct mass measurements with the penning trap setup JYFLTRAP [16] would be of great interest to the community.

Fig. 2. SEM image of ^{232}Th nitrate depositions on a Ti substrate. 432 drops, 20 nL each, of an aqueous solution were positioned using the DoD system. 174 µg of ^{232}Th were deposited in total.

In the case of ^{232}Th, the target material is available in the form of a mechanically stable self-supporting metal foil. For all other, more exotic actinide isotopes, such as ^{229}Th, ^{233}U, ^{239}Pu, or ^{241}Am, this possibility does not exist because there is not enough material available to produce a self-supporting target. Therefore, other production methods must be used to produce thin films on suitable backings. ^{232}Th was selected as the first test species to compare the performance of DoD-based targets with that of ^{232}Th metal foil targets and thus to assess the suitability of printed layers for online experiments.

Numerous ^{232}Th targets with average thicknesses ranging from 197 to 592 µg/cm² were produced. Various metal foils (Pd, Rh, W, Zr, Pt, Ti, Ni, Au), provided by the GSI target laboratory, were tested as possible backing material. The targets were printed from aqueous nitrate solution and subsequently heat-treated at 550°C in a muffle furnace. Fig. 2 shows a SEM image of one of the targets before irradiation. One can see the ^{232}Th-nitrate deposits on the Ti substrate as well-defined spherical structures. However, large uncovered areas are also clearly visible between the deposits. A selection of these targets could be tested during a beam time in 2018 at JYFL. In addition to the mechanical stability, the yield could also be compared to that from a 3.25 mg/cm² ^{232}Th foil.

The results of this beam time were very promising. Reaction products could be obtained from all used DoD targets and all targets withstood the thermal stress in the beam without being visibly damaged. Compared to the solid ^{232}Th film, about 30% of the yield of reaction products was obtained [17]. The yield was presumably limited by insufficient coverage of the irradiated area with target material.

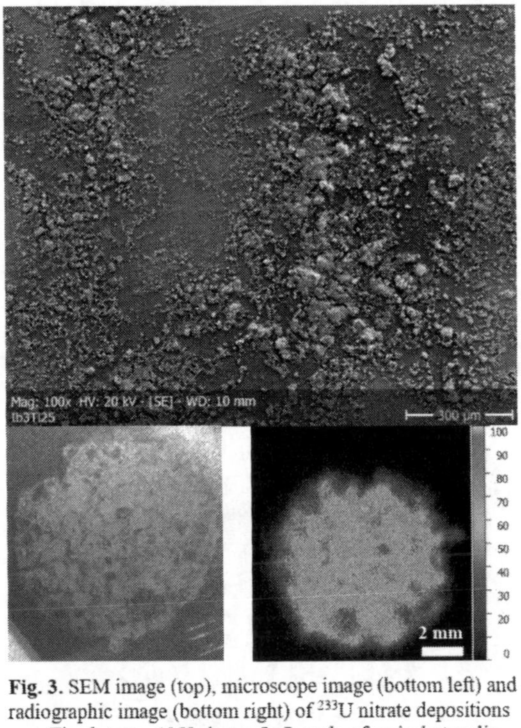

Fig. 3. SEM image (top), microscope image (bottom left) and radiographic image (bottom right) of ^{233}U nitrate depositions on a Ti substrate. 1052 drops, 5 nL each, of an isobutanolic solution were positioned using the DoD system. 61 µg of ^{233}U were deposited in total [18].

In 2019, additional target preparation experiments were performed, this time with ^{233}U as target material, using organic solvents instead of aqueous solutions to achieve better wetting of the droplets on the substrates. Various alcohols, acetone, and acetonitrile were applied as possible solvents. In particular, printing from pure isobutanolic solution led to a more homogeneous distribution of activity than from aqueous solutions. Finally, 6 targets with film thicknesses in the range of 120 to 180 µg/cm² on Ti (25 µm), Pt (20 µm) and Au (20 µm) substrates were sent to Jyväskylä [18]. Originally, experiments with these targets were planned for 2020, but could only be carried out in October 2022.

The ^{233}U targets irradiated in 2022 show a better distribution of the target material on the backing (see fig. 3), although still no completely homogeneous layer could be obtained. However, it can be seen, especially in the radiographic image, that there are only few gaps in the activity distribution where no target material is present at all. In the SEM picture, one can see that the uranyl nitrate formed more small-grained crystallites distributed over the surface.

2.2 Usage of polymer-based solutions

A new idea to increase the homogeneity of the produced layers is the combination of DoD with polymer assisted deposition (PAD). In the PAD process [19,20], a polymer agent is added so that, after drying, a homogeneous polymer film is formed, in which the target material is uniformly distributed. The polymer film is then converted by thermal decomposition, leaving a uniform layer of the target material.

Polyethyleneimine (PEI) was used as the polymer for the initial experiments. PEI offers the advantage that it is readily soluble in water and ethanol and decomposes at relatively low temperatures (< 500°C).

Again, ^{233}U was chosen as the first test species. For this purpose, uranyl nitrate was dissolved in water as well as ethanol and mixed with PEI solutions, which were previously adjusted to a pH in the range of 6.5 - 9.2. Two different PEI solutions were used, one using PEI with a molecular weight of 1800 g/mol and one with 10000 g/mol (Alfa Aesar, 99%). The mixture of a PEI solution with a ^{233}U solution was applied by DoD to the desired substrate (25 µm Ti foils) and then dried under an IR lamp. As a final step, the polymer was decomposed in a muffle furnace at a maximum temperature of 410°C [21].

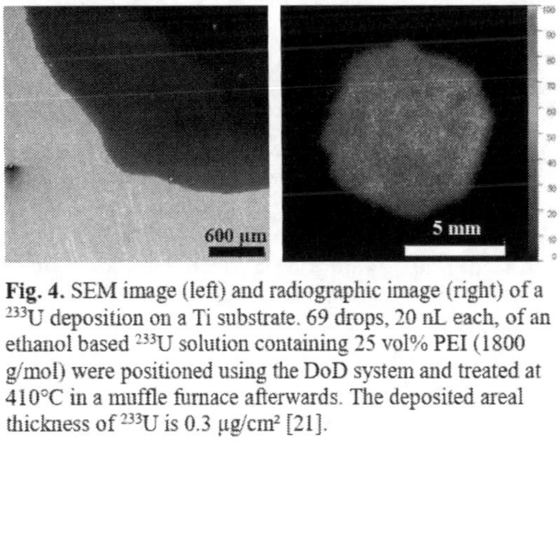

Fig. 4. SEM image (left) and radiographic image (right) of a ^{233}U deposition on a Ti substrate. 69 drops, 20 nL each, of an ethanol based ^{233}U solution containing 25 vol% PEI (1800 g/mol) were positioned using the DoD system and treated at 410°C in a muffle furnace afterwards. The deposited areal thickness of ^{233}U is 0.3 µg/cm² [21].

Figure 4 shows a SEM image (left) and a radiographic image (right) of a deposition containing 0.3 μg/cm² ^{233}U obtained in this way. Compared to all previously discussed DoD methods, an extremely homogeneous layer was obtained here, with no gaps visible in the coating. Also, the original printing pattern is no longer observable because a uniform polymer film formed before the heating procedure.

Thus, this method is extremely promising from the point of view of producing homogeneous, thin layers. However, more optimization steps are necessary before this can be applied. A fundamental problem arose during the actual printing process, as the dispenser tip was regularly clogged due to the high viscosity of the polymer solution, so that it was often not possible to perform a continuous printing sequence. This could be compensated for using the PEI with a molecular weight of 1800 g/mol and using larger dispenser tips (20 - 75 nL), but the printing process remains more error-prone than printing with aqueous or alcoholic solutions.

A second problem arose with the maximum achievable layer thickness. Layers with > 10 μg/cm² could not be achieved so far, because using higher uranium concentrations either changed the printing behavior of the polymer solution, so that a homogeneous film was no longer obtained, or the obtained film no longer adhered to the substrate after the final heating step. There was a tendency to observe that those problems were less pronounced when using the shorter chained PEI. This may be due to incomplete coordination of the dissolved uranyl ions by the PEI, when the uranium concentration is too high. Therefore, an approach for further studies would be to use PEI with even lower molecular weight, but possibly in higher concentration, so that sufficient PEI molecules are available to complex all uranium ions. Possibly, the addition of an additional coordinating agent, e.g., EDTA (ethylenediaminetetraacetic acid), could also positively influence the properties of the polymer film. These ideas will be further investigated in future work.

3 Conclusion & Outlook

The use of organic solvents, like alcohols, acetonitrile, etc., improves the chance of achieving flat deposits and avoiding irregular patterns originating from the surface tension of water droplets, like the so-called coffee ring effect [22].

The best results in terms of uniformity and homogeneity of the deposits have so far been achieved using PEI as polymer-forming agent. So far, this has only been successful up to a low uranium concentration, and the printing process has also been considerably more complex than printing with other solvents due to the properties of PEI. But, there is potential to make further progress by adapting the conditions, for example using PEI with a lower molecular weight but a higher concentration.

Besides changing the solvent, there are other ideas for adjusting the DoD process. For example, chemical or physical pretreatment of the substrate is conceivable. First attempts for anodic oxidation of Ti substrates to improve the wettability of the surface showed promising results [11]. Another possibility to be investigated in the near future is the pretreatment of substrates with a plasma torch, which should also activate the surface so that increased hydrophilicity is achieved.

The DoD technique has great potential to be a complementary production method for thin films of lanthanides and actinides due to its flexibility in the choice of solvents and substrates. Further studies are needed to identify crucial parameters in the production of homogeneous films, but PAD based attempts show promising results.

We acknowledge the support of the technical staff of the TRIGA Mainz research reactor and the corresponding mechanical workshop as well as of the staff of the JYFL accelerator laboratory and Iain Moore for scientific support during the JYFL beam times.

References

1. M. Block et al, Progress in Particle and Nuclear Physics **116**, 103834 (2021)
2. M. Eibach et al, Phys. Rev. C **89**, 064318 (2014)
3. A. Skawran et al, Nuovo Cim. C **42**, 125 (2019)
4. L. von der Wense et al, Nature **533**, 47 (2016)
5. O. Kaleja et al, Phys. Rev. C **106**, 054325 (2022)
6. S. Götz et al, Nucl. Instr. Meth. A **995**, 165090 (2021)
7. Ch. E. Düllmann et al, J. Radioanal. Nucl. Chem., https://doi.org/10.1007/s10967-022-08631-4 (2022)
8. J. Runke et al, J. Radioanal. Nucl. Chem. **299**, 1081 (2014)
9. R. Haas et al, Nucl. Instr. Meth. A **874**, 43 (2017)
10. R. Haas et al, Radiochim. Acta **108**, 923 (2020)
11. Roman Abrosimov, Bachelor thesis, Johannes Gutenberg University Mainz (2020)
12. M. O. Frégeau et al, Nucl. Instr. Meth. A **817**, 35 (2016)
13. Jana Lukacova, Bachelor thesis, Johannes Gutenberg University Mainz (2021)
14. Christopher Sirleaf, student internship report, Johannes Gutenberg University Mainz (2022)
15. J. Äystö, Nuclear Physics A **693**, 477 (2001)
16. T. Eronen et al, Eur. Phys. J. A **48**, 46 (2012)
17. I. Pohjalainen et al, to be submitted
18. Christopher Sirleaf, Bachelor thesis, Johannes Gutenberg University Mainz (2019)
19. A. K. Burrell et al, Chem. Commun. **1271** (2008)
20. M. A. Garcia et al, Nucl. Instr. Meth. A **613**, 396 (2010)
21. Tanyel Schwab, Bachelor thesis, Johannes Gutenberg University Mainz (2021)
22. R. D. Deegan et al, Nature **389**, 827 (1997)

Production of powder targets for neutron-induced cross section measurements

André Moens[1], *Pedro* Amador Celdran[2], *Herwin* Hein[2], *Goedele* Sibbens[1*], *David* Vanleeuw[1], and *Stefaan* Van Winckel[2]

[1]European Commission, Joint Research Centre (JRC), Geel, Belgium
[2]European Commission, Joint Research Centre (JRC), Karlsruhe, Germany

Abstract. Nuclear powder targets for neutron-induced cross section measurements were prepared by pressing. The choice of the production technique was related to the type of nuclear experiment and the quality of the powder. This paper describes the production process of unsintered compacts of ^{94}Mo, ^{95}Mo, ^{96}Mo metal powders and $^{239}PuO_2$ powder with a thickness as low as possible by uniaxial pressing of powders. Special attention goes to the difficulties encountered during the preparation process of $^{239}PuO_2$ pellets. Investigation on the presence of impurities and so the quality of the PuO_2 material was performed by scanning electron microscopy with energy dispersive X-ray spectroscopy, by inductively coupled plasma mass spectrometry, X-ray diffraction and by thermal analysis.

1 Introduction

Targets for neutron-induced cross-section measurements are prepared according to the type of nuclear experiment, the requested target characteristics, final quality of the sample, material and techniques available [1]. In case thick targets are required with an areal density up to g·cm^{-2}, preferably a metallic disc is used. In case the isotope of interest is only available as a powder and the conversion into a metal is not possible, the powder is compacted into mechanically stable pellets [2].

Compaction is a widely applied method. Powder compacts can be obtained by uniaxial pressing, isostatic pressing, metal injection moulding or by less used processes such as powder rolling, extrusion and dynamic or explosive compaction. Often a binder is added to favour densification during the compaction process and a lubrication to decrease friction with the inner wall of the press tool. The compacts are subsequently sintered to induce consolidation and modify the porosity as well as the microstructure irreversibly from contacting particles resulting in compacts with a density higher than 95% of the theoretical density (TD) [3-8].

This paper describes the preparation at JRC Geel of unsintered pellets of ^{94}Mo, ^{95}Mo, ^{96}Mo metal powders and $^{239}PuO2$ powder with a mass of 2 g and with a thickness as low as possible with a maximum of about 1 mm by uniaxial pressing of powders. To avoid addition of impurities and enable simple recovery of limited and valuable material, pellets were produced without adding fluids or binders or lubricants.

The Mo and PuO_2 pellets will be used as targets in neutron-induced cross section measurements. The powder compacts of Mo isotopes are equally important in nuclear astrophysics, nuclear energy application, particularly for GenIV reactors, and in fundamental physics [9, 10]. The production of respective targets is supporting one of the main requests for the H2020 EURATOM Project Supplying Accurate Nuclear Data for energy and non-energy Applications (SANDA) [11]. Neutron-induced cross sections of ^{239}Pu are of great value as ^{239}Pu is one of the "big three" nuclei (^{235}U, ^{238}U, ^{239}Pu) highlighted by the Collaborative International Evaluated Library Organisation (CIELO) Project [12].

The production of the $^{239}PuO_2$ pellets for the measurement of ^{239}Pu (n,n'γ) and (n,2nγ) cross sections within the JRC-wide open access programme EUFRAT [13, 14] got special attention because of mechanical instability and non-homogeneity of the pellets and corrosion of the press tools. For better understanding of these problems, the Pu material was characterized at JRC Karlsruhe for impurities by scanning electron microscopy with energy dispersive X-ray spectroscopy (SEM-EDX), by inductively coupled plasma mass spectrometry (ICP-MS), X-ray diffraction (XRD) and by thermal analysis (TA).

2 Production process

The aim of compaction was to obtain a pellet with sufficient strength to withstand further handling operations without loss of material. The press tool was made of hardened steel with the desired diameter and the corresponding lower and upper dies machined to close tolerances. The maximum filling height and the

* Corresponding author: Goedele.SIBBENS@ec.europa.eu

maximum allowed pressing force, 1.0 kN·mm^{-2} for a diameter above 9.0 mm, was respected in function of the diameter of the die. The press tool was used in combination with the Specac (15-t pressure load) or Retsch (25-t pressure load) press. In case of Pu, the P/O/Weber manual 2-Column Laboratory Press, Model PW 30-BOX, enclosed in a glove box, was used. All consisted of a hydraulic lab press, an external hand pump and had the same standard operation procedures.

The pelletizing process contained three main stages: (1) die filling, (2) compaction and (3) ejection. A quantitative amount of the powder was inserted directly into the matrix. The required pressure force was generated by hand pumping the hydraulic fluid (oil) to raise the piston and reaching the designated press force.

Prior to the production of the 98.97 at% ^{94}Mo, 95.40 at% ^{95}Mo, and 95.90 at% ^{96}Mo metal powder pellets, tests were done with natural Mo metal powder of grain size < 5 µm and of 350 µm, and prior to the $^{239}PuO_2$ pellets CeO_2 and natural UO_2 were used. The tests were performed to find the optimum press parameters for a mechanically stable pellet with a thickness as low as possible and a maximum of about 1 mm. Finally, the Mo pellets were pressed with 2 g of each isotope and a die diameter of 20 mm because the tests with a diameter of 30 mm resulted in broken pellets.

The $^{239}PuO_2$ pellets were produced from different ORNL (Oak Ridge National Laboratory) batches: batch n° 716 (99.97% wt% ^{239}Pu), batch n° 1756 (99.90% wt% ^{239}Pu) and batch n° 1756p (batch n° 1756 after purification to remove Am). Batch n° 1756 originates from the same mother batch as batch n° 716 is, but less enriched in ^{239}Pu. The purification of batch n° 1756 at SCK CEN in Belgium was done by peroxide precipitation, re-dissolution of the plutonium peroxide in nitric acid, oxalate precipitation and calcination for 2h at 735 °C under air. This temperature is above the expected completion of plutonium oxalate decomposition and below the pre-sintering stage. Temperature, atmosphere and hold time were chosen to ensure complete decomposition of the oxalate, yet good compression characteristics of the PuO_2 powder. Sinterability was not a goal, but the conditions were expected to enable good densification [15]. The test with 2 g of natural UO_2 and a diameter of 30 mm showed that the pellet would be too fragile for handling. Therefore, instead of one pellet with a diameter of 30 mm, four $^{239}PuO_2$ pellets with a diameter of 12 mm were produced.

2.1 Characterization

All pellets were characterized for mass by weighing and for thickness with a digital thickness gauge. The density, calculated from the mass, the thickness and the inner diameter of the die, was compared with the TD.

The bad strength of the Pu pellets produced from the first batch n° 716, the heterogeneity of the colour and the corrosion of the press tools raised questions related to the presence of impurities. Therefore, a PuO_2 pellet (n° 050) and part of the original powder of batch n° 716 (n° 051) were analysed by ICP-MS, SEM-EDX, XRD, and differential thermal analysis and thermogravimetric analysis (DTA-GTA).

ICP-MS was done with the 'nuclearised' ElementXR instrument (Thermo Fisher, Bremen, Germany). About 140 mg of the pellet and about 370 mg of the powder were dissolved, each in 30 mL concentrated HNO_3 + 0.1M HF on a hotplate at 90 °C for 48 hr. An important remark is that the samples did not completely dissolve. The supernatant was diluted by a factor of about 10000. The complete mass spectrum was measured from 6 till 248 amu (atomic mass unit).

The elemental composition of the pellet, some particles and the powder were investigated by SEM-EDX at different positions of the material. Analyses were obtained in a Philips/FEI™ XL40 SEM operated at 25 kV, equipped with a SAMx EDX.

The crystal structure was studied by XRD to check the purity of the Pu sample and identify the crystalline phases and composition information. XRD analysis was performed on a Bruker D8 Bragg–Brentano advanced diffractometer (Cu Kα1 radiation) equipped with a Lynxeye linear position sensitive detector and installed inside a glove box under inert atmosphere. The powder diffraction patterns were recorded at room temperature using a step size of 0.01316° with an exposure of 4 s across the angular range $10° \leqslant 2\theta \leqslant 120°$. The operating conditions were 40 kV and 40 mA. Phase analyses were done using the powder diffraction (PDF2) and the inorganic crystal structure database (ICSD). Quantification of the sample was performed by Rietveld refinement using Highscore Plus software.

Thermogravimetry differential TA (TG-DTA) was carried out under flow of oxygen (99.99%) and argon (99.999%) at a heating rate of 10 °C·min^{-1} on a temperature range of Troom till 1400 °C. The apparatus used was NETZSCH Simultaneous Analyzer STA 449 Jupiter. The weight loss of the material was investigated along heating by DTA-GTA to link it to the potential volatile impurities present in the sample.

3 Results and discussion

To produce a pellet with a thickness as low as possible and not exceeding a maximum of 1 mm, the diameter was chosen in function of the amount of material required for the experiments and the compact strength.

Pressing 2 g natural Mo and natural UO_2 with a diameter of 30 mm resulted in pellets with a thickness from 0.5 mm to 0.7 mm. The pellets did not have good strength and broke during the ejection phase. This was not the case for a diameter of 20 mm.

Fig. 1. Picture of a ^{95}Mo pellet with a mass of 1.98 g, a diameter of 20 mm and a thickness of 1.1 mm.

Mechanically stable ^{94}Mo, ^{95}Mo, ^{96}Mo pellets could be prepared with a thickness from 1 mm to 1.1 mm. At the same pressure force of 160 kN, the natural Mo pellets had a thickness of 0.9 mm. The grain size did not influence the pressing process. Fig. 1 shows a picture of a ^{95}Mo pellet with a mass of 1.98 g, a diameter of 20 mm and a thickness of 1.1 mm after ejection on top of the press die. The pellet has a good strength, has sharp borders and glossy surfaces and can be further handled without loss of Mo.

The tests applying the envisaged press parameters were successful with CeO_2 and natural UO_2, resulting in stable pellets of desired diameter of 12 mm and thickness of maximum 1 mm. However, when applying the same press parameters using different $^{239}PuO_2$ materials, this behaviour was not fully reproducible.

The first PuO_2 pellet of batch n° 716 with a diameter of 12 mm could be produced with a good strength while the second, third and fourth pellets lost particulate material. All four had a thickness of about 0.75 mm and showed coloured spots. The press tools that were touching the PuO_2 were corroded after each pressing process. This could be due to impurities in the powder. The press procedure applied to PuO_2 of batch n° 1756 was successful. As shown in Fig. 2a, this material could be pressed into a pellet with a diameter of 12 mm and good strength. Because of the presence of ^{241}Am, the four pellets could not be used for the envisaged experiment. The pellets produced with the PuO_2 purified for Am, batch n° 1756p, showed the same behaviour as the ones from batch n° 716, namely the first pellet had a good strength (Fig.2b) and thereafter the pellets lost particulate material (Fig.2c).

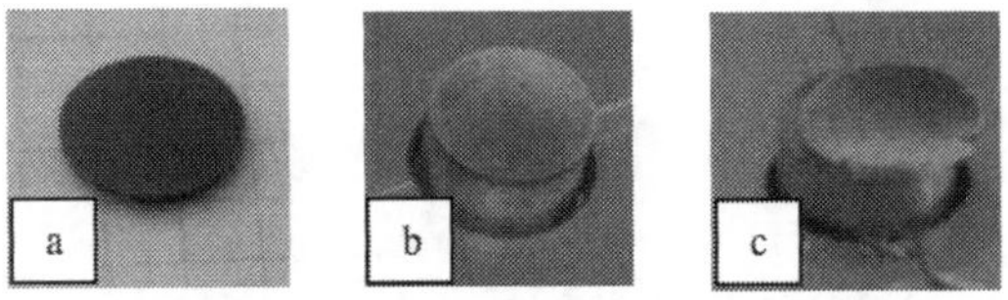

Fig. 2. $^{239}PuO_2$ pellets of batch n° 1756 with a mass of 0.5 g, a diameter of 12 mm and a thickness of about 0.8 mm prepared before (a), and after (b) (c) the purification process for Am, (b) is the first pellet, (c) the second pellet prepared.

In addition, the press tools that were touching the PuO_2 were corroded after each pressing process. The observed variability in the colour of the PuO_2 is well known, and while PuO_2 is normally olive green, samples of yellow, buff, khaki, tan, slate, and black are also common. It is generally believed that the colour is a function of particle size, chemical purity, stoichiometry, and method of preparation although the colour of PuO_2 resulting from a given method of preparation is not always reproducible [16].

The concentrations of impurities in the Pu batch n° 716 measured by ICP-MS are presented in Table 1 and expressed in µg/g of the original solid material. As the dissolution was not complete, results should be considered as indicative. Other lanthanides could also be seen at trace level (few ppm). The actinide elements come from the radioactive decay of plutonium. The Ce could be related to the tests with CeO_2 in the glove box. The average concentrations of impurities in Pu batch n° 716 showed also V, which was confirmed by SEM-EDX.

Table 1. Average concentrations of impurities in Pu batch n° 716 in µg/g original solid material determined by ICP-MS

V	U	Ce	Np	Th	Ta
µg/g solid material					
12400	3340	20	15	9	5

The EDX analyses showed the presence of several elements in Pu batch n° 716. Table 2 presents the weight percentages in the powder at positions 12-13-14-15 of the SEM image (Fig. 3). The SEM image showed a morphology of irregular crystals, rosettes, and thin plates of different sizes.

Table 2. EDX analyses at position 12-13-14-15 of Fig. 3

pos	C wt%	O wt%	Si wt%	Cl wt%
12	2.93	14.37	1.49	2.04
13	4.06	13.77	0.65	1.69
14	13.36	10.58	0.58	1.10
15	6.00	12.66	1.06	1.68

pos	Ca wt%	Fe wt%	U wt%	Pu wt%
12	1.48	0.55	2.92	74.23
13	1.82	0.23	2.81	74.96
14	1.34	0.44	1.95	70.65
15	1.54	0.33	2.28	74.46

A review of the literature suggested that the morphology of PuO_2 formed through the Pu(IV) oxalate process can differ based on the parameters and conditions chosen during sample preparation [17, 18]. The presence of C could be linked to a thermal decomposition process of the hydrated Pu oxalate $Pu(C_2O_4)_2 \cdot 6H_2O$ to produce PuO_2 powder. This process results in the oxalate ligand decomposition and production of amorphous Pu oxycarbide species, CO_2, and ultimately crystalline PuO_2 with some residual C-containing species [17]. This would also explain the colour of the powder, which was black and light brown. If the decomposition would have been successful, the expected colour of the PuO_2 powder would have been homogeneous black [18].

Fig. 3. SEM powder position 12-13-14-15

The XRD analysis showed that the sample was not homogeneous and pure. It contained at least 2 phases with PuO_2 as the main component. The peak shape, broad at bottom and sharp at the top relates to the heterogeneity in the crystallinity of the sample. This indicated an amorphous part. Based on a phase search in the ICSD and PDF2 databases using the elements and element family of Pu, O, C, Cl and Np as reported in the ICP-MS and SEM-EDX, the second phase could be assigned to a compound isostructural to $H_2F_{12}Np_3O$ [98-004-7190] with a monoclinic structure. However, a compound isostructural to $PuF_4(H_2O)1.6$ [01-074-1766] with a cubic structure could not be excluded. This second phase, present at about 15 wt% in the sample, included a halogen element. F was given because this element was known in the database of phases. It could also be Cl instead of F as $PuCl_4$ has the same structure as PuF_4 and Cl was also confirmed by EDX analyses.

The DTA-GTA displayed about 11 wt% weight loss for PuO_2 powder in an O_2 and in an Ar atmosphere. Along the temperature, the weight decreased regularly till about 700 °C and then along two steps, [700-800] °C and [800-1100] °C. The pellet showed a weight loss of 9% under O_2, while for Ar the final step was not reached yet at 1400 °C. As expected, the reaction was stronger with the powder than with the pellet due to the larger surface available for the reaction. The weight loss took place along three steps: 1) <700 °C; 2) >700 °C and >800 °C; and 3) >800 °C. The relatively important weight loss of 11 wt% as a result of the DTA-GTA on the Pu batch n° 716 showed that the sample could not be pure PuO_2. By comparison, for pure Pu-oxalate, a weight loss of about 35 wt% is expected.

4 Conclusion

Several pellets of stable Mo metal powder and of $^{239}PuO_2$ powder were prepared as targets for neutron-induced cross section measurements within SANDA [11] and EUFRAT [14], respectively. The pellets were produced by uniaxial pressing the powders with a hydraulic press, without adding fluids, additives or lubricants. The pressure force and the diameter of the pellets were optimized to produce mechanically stable unsintered pellets with a mass of 2 g and a thickness as thin as possible and maximum 1 mm. The PuO_2 pellets made of PuO_2 of batch n° 716 were characterised by ICP-MS, SEM-EDX, XRD and DTA-GTA.

The Mo pellets had good strength, sharp borders and glossy surfaces and could be further handled without loss of material. The compaction of the PuO_2 powder however resulted in bad strength quality for the purified batches. Only pellets from the Pu batch unpurified for Am could be produced with a good strength.

Powder contamination can influence the pressing process. The amount and the type of impurities depend on the powder production process, the storage container, the purification process, the process environment, etc., which is not always known. Crystalline order and morphology of PuO_2 grains and aggregates are known to be sensitive to the initial production conditions [19]. The analyses of the PuO_2 of batch n° 716 by ICP-MS, SEM-EDX, XRD and DTA-GTA showed beside the expected Pu, O and actinide elements from the radioactive decay of Pu, the important presence of V, C, and Cl. C could be linked to the purification process of PuO_2 via thermal decomposition of hydrated Pu(IV) oxalate. We can only assume that the presence of C, CO_2, and Cl in the purified Pu, in combination with water absorbed from the environment, caused corrosion of the steel press tools. This could result in end-capping by friction of the Pu with the inner wall of the die sleeve and/or that volatile impurities caused high porosity and cracks in the Pu pellet. A heat treatment of the Pu material under oxidizing atmosphere would favour the purification of the material. It would desorb and decompose impurities formed by the moisture, potential presence of chloride and rest of oxalate if any.

The authors would like to thank M. Ernstberger for the SEM-EDX and E. Dahms for the XRD (JRC Karlsruhe). Also for their collaboration and fruitful discussion, the authors want to acknowledge O. Walter and R. Eloirdi (JRC Karlsruhe) and M. Verwerft (SCK CEN, Belgium).

References

1. G. Sibbens, et al., EPJ Web Conf. **229**, 04003 (2020)
2. V. Verdingh, NIM **102**, 431 (1972)
3. F. Lemoisson, L. Froyen, Fundamentals of Metallurgy, Book chapter 1555602 - Understanding and improving powder metallurgical processes, **471** (2005)
4. L.F. Francis, Materials Processing, Chapter 5 - Powder Processes, **343** (2016)
5. V. Tyrpekl, et al., Rev. Sci. Instrum. **86**, 023904 (2015)
6. P. Balakrishna, et al., IJEMS **5**, 136 (1998)
7. A.I. Bobylev, et al., At. Energy **96**, 65 (2004)
8. R.J. Vedder, Preparation of Sintered $^{237}NpO_2$ Pellets for Irradiation to Produce ^{238}Pu Oxide, ORNL/TM-2017/728 (2018)
9. R. Mucciola, Il Nuovo Cimento **44 C**, 62 (2021)
10. R. Mucciola, et al., NIM B **531**, 100 (2022)
11. http://www.sanda-nd.eu/
12. M. Fleming, et al., EPJ Web Conf. **239**, 15003 (2020)
13. M. Kerveno, et al., Eur. Phys. J. A **51**, 167 (2015)
14. https://joint-research-centre.ec.europa.eu/tools-and-laboratories/open-access-jrc-research-infrastructures_en
15. K. Verguts, K. Van Hecke, Purification of an Am-241 contaminated Pu-239 batch, Report R-8843, SCK CEN, Belgium (2022)
16. Nuclear Material Research and Technology, Los Alamos National Laboratory, 2nd quarter (2004)
17. R.M. Orr, et al., J. Nucl. Mater. **465**, 756 (2015)
18. J.H. Christian, et al., J. Nucl. Mater. **562**, 153574 (2022)
19. E. Villa-Aleman, et al., Journal of Nuclear Materials **515**, 140 (2019)

Application of multiple source preparation procedures on environmental samples for determination of Uranium in Alpha Spectrometry

Maruf Abubakar[1*], *Martina* Rozmaric[2], and *Ljudmila* Benedik [3]

[1]Ghana Atomic Energy Commission, Radiation Protection Institute, Accra, Ghana
[2]Environmental Protection Agency, Office of Radiation Protection and Environmental Monitoring, Dublin, Ireland
[3]Jožef Stefan Institute, Ljubljana, Slovenia

Abstract. Different source preparation procedures for determination of uranium radioisotopes in environmental samples by alpha spectrometry are presented. Addition of ^{232}U tracer was followed by the preconcentration of radionuclides of interest from the samples by co-precipitation on $Fe(OH)_3$ at pH 9 using ammonia solution. The uranium extracts were obtained by column separation using UTEVA resin after which the uranium sources from the samples were prepared using either electrodeposition or micro-coprecipitation with LaF_3 before counting. The procedures were tested on spinach, soil, tap water and waste water from the Libanon Gold mine and the results obtained were compared.

1 Introduction

Alpha-particle spectrometry is a measurement technique that has found many practical applications in such diverse fields as nuclear decay data measurements, geological studies, or the measurement of low levels of activity in the environment.[1] The samples used in this study are water, spinach and soil. Complete decomposition of the spinach and soil sample matrix is necessary for an accurate radiochemical analysis of alpha-emitting radionuclides. Therefore, conventional wet digestion was used to achieve this. The aim of the work presented is to investigate two different source preparation approaches for determination of low-level uranium radioisotopes in water, spinach and soil samples by alpha-particle spectrometry. The preconcentration method for uranium was used by co-precipitation with $Fe(OH)_3$ at pH 9 using an ammonia solution in the water samples.[2] Micro-coprecipitation with LaF_3 method was used for the tap water and waste water from the mine while electrodeposition was used for the soil and spinach samples after uranium separation using a UTEVA column. These methods allow the radionuclides of interest and the tracers to be made into a thin, flat and uniform source, which allows adequate transmission of the alpha-particles to the surface of the detector.[2]

2 Methodology

2.1 Spinach

A triplicate of 2 g of ashed spinach sample was weighed into a glass beaker and 10 mL of concentrated HNO_3 was added to make a slurry. 0.5 g of ^{232}U tracer of concentration 0.3163 Bq/g was added and heated for three hours at 150°C. 1 mL H_2O_2 was added into the sample and evaporated to dryness. 6 mL of concentrated HNO_3 and 1 mL of H_2O_2 (6% w/v) were added into the sample and evaporated to dryness at 250 °C[1]. The sample was dissolved in 3 M HNO_3 and loaded on a chromatographic column of length 10 cm containing UTEVA resin (Uranium and TEtraValents Actinides) of size 100 µg. The extractant for the anion exchange chromatography in the UTEVA Resin is diamyl, amylphosphonate (DAAP), which extracts nitrato complexes of the actinide elements at room temperature. The container is rinsed once with 5 mL 3 M HNO_3 and the washing solution is also loaded on the column. The column was washed with 20 mL 3 M HNO_3, 5 mL 9 M HCl, and 25 mL 5 M HCl after which uranium was eluted with 15 mL 1 M HCl into a clean tube.[5] The eluted solution was evaporated to dryness. 1 mL of concentrated HNO_3 and 1 mL of H_2O_2 were added and evaporated to dryness.[4] The residue was dissolved in 1 mL of 6 M HCl and then transferred into the electrodeposition cell shown in figure 1. The electrodeposition method was a modified Talvitie's and Hallstadius's method whereby the chemically separated uranium was further dissolved in 10 mL of electrolytic solution made up of 5.7% Ammonium oxalate in 0.3M

* Corresponding author: mcrufguy@yahoo.com

HCl.[6] The deposition planchet which served as the anode in the electrodeposition setup is a stainless steel disk of diameter 25 mm. The electrodeposition was started with an electrical current of 0.6 A with an initial voltage of 9.26 V fluctuating to a maximum voltage of 9.59 V for 2 hours at room temperature. 1 mL of ammonia was added a minute before putting off the power supply. The deposition diameter was 20 mm. The disc with electrodeposited radionuclides was dried and counted.

2.2 Soil:

A triplicate of 2 g of oven-dried soil sample was weighed into a glass beaker and 10 mL of concentrated HNO_3 was added to make a slurry. 0.5 g of ^{232}U tracer of concentration 0.3163Bq/g was added and heated for three hours at 150°C. 10 ml each of 70% w/w concentrations of HNO_3, HF, $HClO_4$ were added to each of the samples and evaporated to dryness on a hot plate. 2 mL of concentrated HNO_3 was added into the sample and evaporate to dryness. The sample was dissolved in 3 M HNO_3 and loaded on the column containing UTEVA resin. The container is rinsed once with 5 mL 3 M HNO_3 and loaded on a column. The column is washed with 20 mL 3 M HNO_3, 5 mL 9 M HCl, and 25 mL 5 M HCl after which uranium was eluted with 15 mL 1 M HCl into a clean tube. The eluted solution was evaporated to dryness. 1 mL of concentrated HNO_3 and 1 mL of H_2O_2 were added and evaporated to dryness. The residue was dissolved in 1 mL of 6 M HCl and then transferred into the electrodeposition cell. The electrodeposition was started with an electrical current of 0.6 A for 2 hours. 1 mL of ammonia was added a minute before putting off the power supply. The disc with electrodeposited radionuclides was dried and counted.

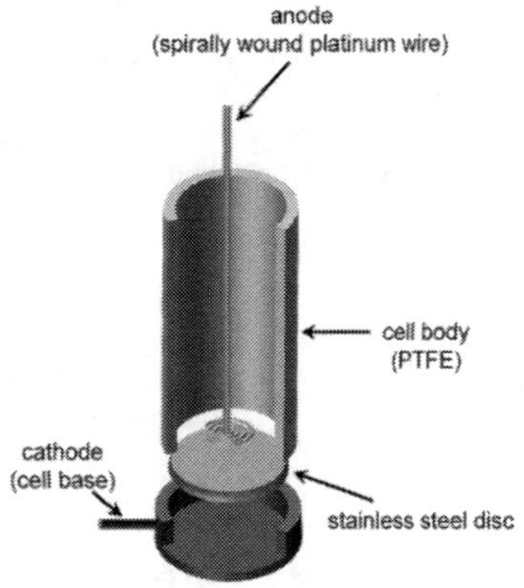

Fig. 1. Schematic diagram of the electrodeposition cell.[6]

2.3 Water:

Three samples each of volume 4 L of tap water and three samples each of volume 100 mL of waste water from the mine were measured into different beakers. Each of the samples was acidified with 12 mL of concentrated HNO_3 after which 0.5 g of ^{232}U tracer was added into each sample. 8 mL of Fe^{3+} carrier (5 mg of Fe/mL) was added to every sample to initiate the co-precipitation of the radionuclides of interest with $Fe(OH)_3$.[4] The pH of the samples was changed to 9.6 by gradually adding concentrated NH_3. Each solution was stirred for 3 hours and allowed to settle down by leaving them overnight. The supernatant was then decanted, centrifuged at 3500 rpm for 5 minutes and rinsed with deionised water until neutral pH was obtained. The $Fe(OH)_3$ precipitate of each was then dissolved in 5 mL of 3 M HNO_3. Preconditioning of UTEVA column was done by rinsing the column with 10 mL H_2O, 10 mL 1M HNO_3 and 10 mL 3 M HNO_3. Each sample was loaded on the column. The centrifuge tubes were rinsed twice with 5 mL 3 M HNO_3 and loaded on every column after which the column was rinsed with 20 mL 3 M HNO_3, 5 mL 9 M HCl, 25 mL 5 M HCl. Uranium was finally eluted with 15 mL of 1 M HCl into a clean tube. 0.1 mL of lanthanum tracer La^{3+} (1 mg/mL), 1 mL $TiCl_3$ and 1 mL concentrated HF were added in the uranium fraction. Each of the samples was left for a minimum of 30 minutes in an ice bath. An Eichrom polypropylene filter disk of diameter 25mm and size 0.1μ size was placed on a funnel for vacuum filtration after which it was rinsed with deionised water and 10 mL of LaF_3 suspension of 0.2 mg/mL.[5] The sample was passed through the filter paper and filtrated.[2] The tube was rinsed twice with 5 mL 0.58 M HF, deionised water and filtrated. Each filter paper was then removed from the funnel with a pair of tweezers to avoid curling, dried under an IR lamp and stuck onto aluminium disk for counting.

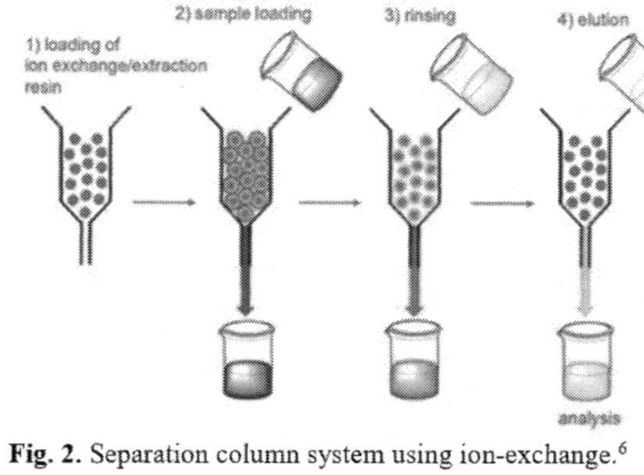

Fig. 2. Separation column system using ion-exchange.[6]

Counting:

The measurements were done using 450 mm^2 active surface Passivated Implanted Planar Silicon (PIPS) semiconductor detectors installed in the 12-chamber Alpha Analyst System (Canberra). The measurements were carried out at a source to detector distance of about 5 cm. The accumulation and analysis of Alpha spectra was done using Genie 2000 software with measurement time of about 144,000s. The calibration of the detectors was made with a standard radionuclide source containing ^{238}U, ^{234}U, and ^{239}Pu, ^{241}Am. The yielded Alpha counting efficiencies are shown in Table 1, with Mine Water samples (MW), Tap Water samples (TW), Spinach samples (SP) and Soil samples (SL).

Table 1. Counting efficiencies and radiochemical recovery for all samples

	U-238	U-235	U-234	Radiochemical Recovery (%)
TW-1	*0.128*	*0.128*	*0.128*	*64.6*
TW-2	*0.133*	*0.133*	*0.133*	*54.5*
TW-3	*0.132*	*0.132*	*0.132*	*59.5*
MW-1	*0.126*	*0.126*	*0.126*	*83.6*
MW-2	*0.130*	*0.130*	*0.130*	*76.1*
MW-3	*0.124*	*0.124*	*0.124*	*72.5*
SP-1	*0.133*	*0.133*	*0.133*	*77.4*
SP-2	*0.132*	*0.132*	*0.132*	*64.8*
SP-3	*0.129*	*0.129*	*0.129*	*51.9*
SL-1	*0.130*	*0.130*	*0.130*	*71.1*
SL-2	*0.130*	*0.130*	*0.130*	*77.7*
SL-3	*0.135*	*0.135*	*0.135*	*74.4*

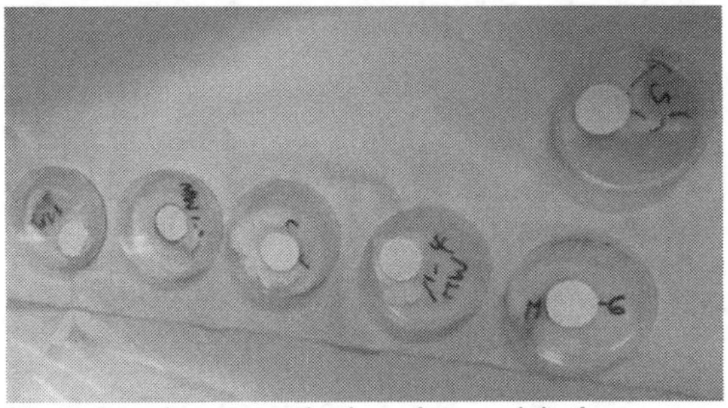

Fig. 3. Samples prepared using microprecipitation procedure ready for counting.

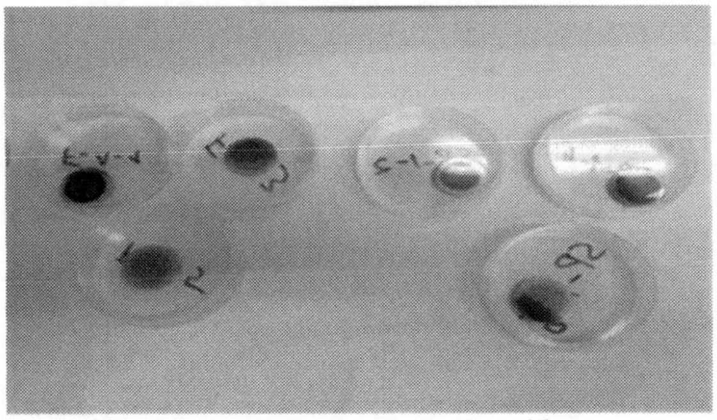

Fig. 4. Samples prepared using electrodeposition procedure ready for counting.

3 Results and discussion

After counting for about 144,000s to obtain enough counts for both the background and the samples, the activity concentrations of all the samples were obtained. The activity concentrations of the Tap Water, Mine Water, Spinach and Soil samples are shown in the graphs below:

Table 2. Activity Concentrations of ^{238}U, ^{235}U and ^{234}U in Tap Water

	TW-1	TW-2	TW-3
U-238 (mBq/L)	**1.28**	**1.55**	**1.6**
Expanded Uncertainty (k=2)	*0.34*	*0.39*	*0.42*
U-235 (mBq/L)	**0.36**	**0.75**	**0.36**
Expanded Uncertainty (k=2)	*0.18*	*0.27*	*0.19*
U-234 (mBq/L)	**3.67**	**3.71**	**4.12**
Expanded Uncertainty (k=2)	*0.63*	*0.66*	*0.74*

Table 3. Activity Concentrations of ^{238}U, ^{235}U and ^{234}U in Mine Water

	MW-1	MW-2	MW-3
U-238 (mBq/L)	**1017.78**	**959.85**	**1093.46**
Expanded Uncertainty (k=2)	*67.51*	*100.67*	*111.93*
U-235 (mBq/L)	**153.98**	**282.16**	**239.83**
Expanded Uncertainty (k=2)	*27.9*	*43.35*	*40.22*
U-234 (mBq/L)	**995.94**	**910.73**	**1154.44**
Expanded Uncertainty (k=2)	*93.75*	*96.74*	*116.68*

Table 4. Activity Concentrations of ^{238}U, ^{235}U and ^{234}U in Spinach

	SP-1	SP-2	SP-3
U-238 (Bq/kg)	**1.18**	**1.28**	**1.55**
Expanded Uncertainty (k=2)	*0.3*	*0.29*	*0.54*
U-235 (Bq/kg)	**0.56**	**0.6**	**1.12**
Expanded Uncertainty (k=2)	*0.2*	*0.2*	*0.45*
U-234 (Bq/kg)	**3.76**	**3.43**	**4.02**
Expanded Uncertainty (k=2)	*0.6*	*0.53*	*0.96*

Table 5. Activity Concentrations of ^{238}U, ^{235}U and ^{234}U in Soil

	SL-1	SL-2	SL-3
U-238 (Bq/kg)	**30.97**	**32.96**	**30.43**
Expanded Uncertainty (k=2)	*10.52*	*10.77*	*9.86*
U-235 (Bq/kg)	**6.88**	**2.2**	**3.93**
Expanded Uncertainty (k=2)	*4.21*	*2.24*	*2.89*
U-234 (Bq/kg)	**26.96**	**26.37**	**24.54**
Expanded Uncertainty (k=2)	*9.59*	*9.26*	*8.52*

The results show the activity concentrations of the mine water samples being greater than the activity concentrations of the tap water samples as predicted. In the first set of samples (MW), ^{238}U and ^{234}U have similar activity concentrations with ^{235}U having the lowest as expected. In the second set of samples (TW) the activity concentration of ^{238}U is almost half of that of ^{234}U with ^{235}U having the lowest concentration.

The results of the second set of data show the activity concentrations of the soil samples being greater than the activity concentrations of the spinach samples as predicted. With regards to the activity concentrations of the soil samples, ^{238}U and ^{234}U have similar activity concentrations with ^{235}U having the lowest. However, the activity concentration of ^{234}U in the spinach samples is almost three times that of ^{238}U with ^{235}U having the lowest concentration.

The radiochemical recovery of the individual sample (see Table 1) is defined as

$$Y(\%) = \frac{net\ P(^{232}U)}{t \times m(^{232}U) \times A(^{232}U) \times \varepsilon} \times 100\%$$

where:

$Y(\%)$ = radiochemical recovery
$P(^{232}U)$ = net count for ^{232}U
$m(^{232}U)$ = net count for ^{232}U
$A(^{232}U)$ = current tracer activity
ε = counting efficiency for tracer

Good radiochemical recoveries were obtained for this particular method with average values of 74.4% (MW), 59.5% (TW), 64.7% (SP) and 77.4% (SL). It is well-known that soil samples are often considered to be one of the most difficult matrices to digest. The Soil average radiochemical recovery of 77.4% being the highest indicates that both the digestion and the source preparation procedures were very effective and points to the fact that this method also works for samples with matrices that are tough to digest.

4 Conclusion

This work presents two of the most effective ways of source preparation; micro co-precipitation technique with LaF_3 following column separation using UTEVA resin for two types of water samples; tap water and waste water from the mine, and electrodeposition for spinach and soil samples. The results obtained show that the method allows for a fast and efficient determination of uranium activity concentrations in all of the samples analyzed.

References

1. Selvig L. K., Inn K. G. W., Outola I. M. J., Kurosaki H. K., Lee A. J., Radioanal. Nucl. Chem. 2005, 263, 341−348.
2. Trdin M., Necemer M., Benedik L., Anal. Chem. 2017, 89, 3169−3176.
3. Inn K. G. W., Robin-Hutchinson J. M., Kelly W. R., Greenberg R., Norris A., Krey P., Feiner M. S., Fisenne E., Popplewell D. S., Gladney E., Beasley T., Huh C. A., Percival D. R., J. Radioanal. Nucl. Chem. 2016, 307, 2513−2520.
4. Jurecic S., Benedik L., Planinsek P., Necemer M., Kump P., Pihlar B., Appl. Radiat. Isot. 2014, 87, 61−65.
5. Eichrom Technologies, Inc. Uranium and Thorium in water; Analytical Procedures ACW01 Rev. 1.7; Eichrom Technologies Inc., Lisle, IL, 2001.
6. Ko YG., Lim J-M., Choi G-S., Chung H., Kang J., (2015). Colloid Surf A Physicochem Eng Asp 487:121–130

Method for releasing carbon foils from substrates

Hiroo Hasebe[1]*, *Hiroki* Okuno[1], *Hiroshi* Imao[1], *Nobuhisa* Fukunishi[1], *Masayuki* Kase[1], and *Osamu* Kamigaito[1]

[1]Nishina Center for Accelerator-Based Science, RIKEN, 2-1 Hirosawa, Wako, Saitama 351-0198, Japan

Abstract. The performance of a carbon foil (C-foil) as a charge stripper in an accelerator is directly affected by its condition and quality. C-foils are fabricated by evaporating carbon onto a substrate. Therefore, a suitable technique must be established for depositing and removing C-foils from substrates. The selection of appropriate releasing agents is crucial for the removal process. C-foils with thicknesses less than 0.1 mg/cm^2 were produced using the arc discharge method, wherein chloride materials were used as releasing agents. The magnetron sputtering method is suitable for manufacturing C-foils thicker than 0.1 mg/cm^2 using appropriate releasing agents, such as fluorine-based or silicon-based agents. The contact angle analysis demonstrates a negative relationship between the C-foil thickness and the wettability of the substrate treated with an appropriate releasing agent. The possibility of using ionic liquids as novel releasing agents is also investigated.

1 Introduction

The development of long-life carbon foils (C-foils) for charge strippers at RIKEN RI Beam Factory began in 1999. Because C-foils are manufactured via vapor deposition on a glass substrate or silicon wafer, they must be removed from the substrate or wafer prior to their installation in the accelerator beam line. The condition and quality of the C-foils directly affect the C-foil lifetime and stability of the charge-stripped beam intensity. Therefore, a suitable technique is required for removing C-foils from the substrates.

The selection of an optimized releasing agent, which is applied on the substrate before the evaporation step, is crucial for the removal process. This agent aids the removal of the carbon layer, maintaining the C-foil condition as intact as possible [1]. Generally, the carbon layer is removed such that the foil floats easily on water. However, humidity entering carbon layer itself negatively affects the foil removal and causes damage resulting in a foil breakage. Therefore, it is necessary to devise a method to prevent humidity from entering the carbon layer while allowing water to permeate between the carbon layer and the substrate at an appropriate rate. This is the reason why we usually use hydrophilic releasing agents such as chloride, soap (a surfactant), and betaine saccharose. However, unless releasing agents with optimum amount and thickness are used, especially in the case of large thickness, water rapidly penetrates into the carbon layer and the C-foil breaks immediately. On the other hand, if the amount of the releasing agent is too small, the carbon foil would not be removed at all.

2 Releasing agents for the arc discharge method

A thin C-foil with a thickness ranging from 10-80 μg/cm^2 was fabricated using the arc discharge method [2]. Before evaporating carbon, a chloride-based releasing agent was deposited onto the substrate using a resistance-heated source. Here, powder-type $NiCl_2$ and $LaCl_3$ were used as releasing agents because powder-type agents had smaller particles and were easier to deposit than granular type. The thickness of the releasing agent was measured using a crystal rate thickness monitor and was equivalent to 10 μg/cm^2 of carbon. The C-foil was removed from the substrate using the general method of floatation on water and was subsequently attached to a holder.

3 Releasing agents for the magnetron sputtering method

A C-foil thicker than 0.1 mg/cm^2 was fabricated using the magnetron sputtering method [3]. Similar to the case of the thin C-foils, the substrate surface was initially treated with a releasing agent. After testing various releasing agents manufactured by different companies, it was determined that using different releasing agents based on the thickness range is optimal. For carbon thicknesses of 0.1–0.3 mg/cm^2, fluorine-based releasing agents, such as DAIFREE (GA-6310) [4] are suitable. For C-foil thicker than 0.3 mg/cm^2, a silicon-based releasing agent RELEASE [5] is suitable.
The substrate surface was treated as follows:

* Corresponding author: hasebe@riken.jp

1. The releasing agent was uniformly sprayed onto the substrate using spraying cans, and thereafter, the substrate was dried.
2. The agent on the surface was wiped off with low-dust paper or cloth.
3. The agent layer was wiped thoroughly until it was not visible anymore to the naked eye.

This is the general procedure for applying releasing agents. It takes several hours to overnight for a C-foil to float on the water while it takes only a few minutes in the case of the C-foil by the arc discharge method using the chloride-releasing agent. Most of the time is spent for water to penetrate between the carbon layer and the substrate. Our tests showed that this releasing agent can be applied to the fabrication of thin metal foils, such as palladium foils.

4 Contact angle measurements for various substrates

4.1 Method for measuring contact angles

To quantify the surface condition of the substrate before C-foil deposition, the contact angles of pure water were measured on various substrates using the sessile drop method. The apparatus used was the KYOWA Interface Measurement and Analysis System (FAMAS). Two microliters of pure water were dropped on the substrate, and after 5 seconds, the contact angle was measured using FAMAS.

4.2 Contact angle on the glass slide

The contact angles of pure water on a glass slide were measured with and without the application of a releasing agent. Figure 1 shows photographs of the contact angle measurements using glass slides treated with chloride-based agents: (a) $NiCl_2$ and (b) $LaCl_3$. The contact angle on the untreated glass slide was 20.2°. The angles measured for the samples treated with $NiCl_2$ and $LaCl_3$ were 31.5° and 36.5°, respectively. The contact angles measured on the slides treated with chloride agents were larger than those on the untreated slides, indicating a decrease in wettability. The contact angle was larger for substrate treated with $LaCl_3$ than that with $NiCl_2$.

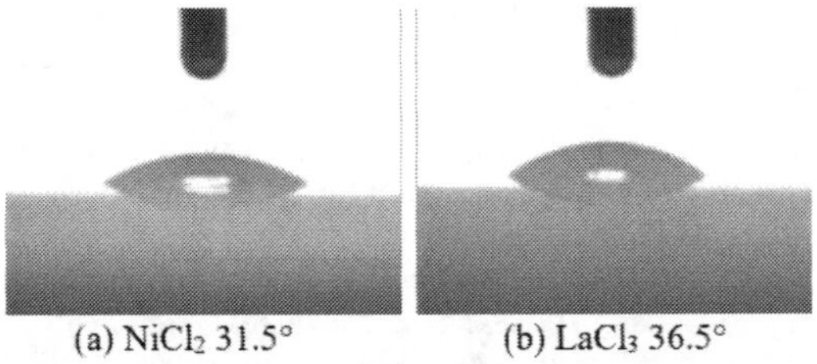

(a) $NiCl_2$ 31.5° (b) $LaCl_3$ 36.5°

Fig. 1. Droplets on glass slides treated with chloride-based releasing agents: (a) $NiCl_2$ and (b) $LaCl_3$. Contact angle in each case is provided.

4.3 Contact angles on the Si wafer

The contact angles on a Si wafer were measured after treating it with fluorine- and silicon-based releasing agents. Figure 2 shows photographs of the measurements obtained using the Si wafer. Figure 2 (a) and (b) show the contact angles of the droplets on wafers treated with the fluorine-based agent DAIFREE and the silicon-based agent RELEASE, respectively. The contact angles on the untreated Si wafers and Si wafers treated with DAIFREE and RELEASE are 30.2°, 57.6°, and 103.5°, respectively. Similar to the case of the glass substrate, the contact angles were larger for the treated wafers than those for the untreated ones. Furthermore, the application of silicon-based RELEASE led to significantly larger contact angles than that of fluorine-based DAIFREE.

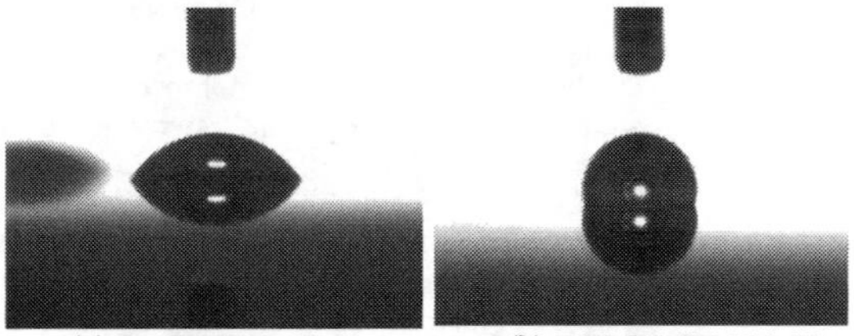

(a) DAIFREE 57.6° (b) RELEASE 103.5°

Fig. 2. Droplets on the Si wafers treated with (a) fluorine-based agent DAIFREE and (b) silicon-based agent RELEASE. Contact angle in each case is also described.

4.4 Summary of contact angles

Table 1 lists the measured contact angles for each combination of the substrate and releasing agent along with the C-foil thickness ranges and deposition methods. Release agents giving higher contact angles are needed for foils with increased thickness. In other words, a negative relationship exists between the C-foil thickness and the required wettability of the substrate.

Table 1. Summary of the contact angles and C-foil thicknesses

Thickness (μg/cm²)	Method	Releasing agent	Substrate	Contact angle (deg)
10-40	AC arc	$NiCl_2$	Slide glass	31.5
60-80	AC+DC +AC	$LaCl_3$	Slide glass	36.5
100-300	Magnetron sputtering	DAIFREE Fluorine based	5-inch Si wafer	57.6
300-10000	Magnetron sputtering	RELEASE Silicon based	5-inch Si wafer	103.5

5 Other releasing agents

Furthermore, potential candidates for application as efficient releasing agents have also been identified. Ionic liquids (ILs), which negligibly evaporate at room temperature, or even in vacuum, are potentially promising materials as releasing agents. Since ILs do not easily change state in any environment, it is thought that the effect on the C-foil is small as well even if it is taken out of the chamber after vapor deposition. Moreover, if IL is deposited on a substrate while maintaining its properties as a liquid, an extremely thin and uniform layer of release agent can be formed.

This time, a water-soluble IL (ILA48-32) [6] was tested to determine the feasibility of using it as a releasing agent. This IL had a decomposition temperature of 170 °C. However, most ILs are toxic and require careful handling.

The IL was tested as a releasing agent as follows:

1. The IL was placed in a deep Ta boat and covered with a perforated Ta lid (Fig. 3 (a)).
2. The IL was evaporated with a resistance-heating evaporation source.
3. Carbon was deposited via the arc discharge method.
4. The carbon-deposited substrate was removed from the device and allowed to float on water. The carbon layer was then peeled off.

The peeled C-foil condition is poor, as shown in Fig. 3 (b). This might be because the IL amount was not optimized, and the IL did not spread uniformly on the substrate. The optimization of the IL amount is difficult because the thickness of the IL layer cannot be measured using a crystal rate thickness monitor. Herein, the entire releasing agent is heated simultaneously, and the IL is immediately boiled, forming various sizes of molecular clusters of the IL as a flux to the substrate. As a result, the IL does not adhere uniformly to the substrate because of the various sizes of the molecular clusters in the deposits.

It is thought that when the IL amount is controlled and a small amount of IL is evaporated many times in a short period to result in a flux with a uniform molecular density, the IL will adhere uniformly to the substrate. However, the evaporation source with a Ta boat used in this study is not suitable for this purpose. The laser evaporation of IL would be a better method for C-foil fabrication compared to that used in this study.

(a) (b)

Fig. 3. (a) IL in the deep Ta boat and its cover. (b) C-foil removed from the IL-treated substrate and floated on water.

6 Summary

Chloride-based release agents are suitable for C-foils with thicknesses of 10–80 μg/cm², which are fabricated using the arc discharge method. Among the chloride materials, $NiCl_2$ and $LaCl_3$ are suitable for C-foil thicknesses lower and higher than 50 μg/cm², respectively. Fluorine- and silicon-based releasing agents are appropriate for C-foil thicknesses ranging from 0.1–0.3 and 0.3–10 mg/cm², respectively. The damage caused by humidity is significantly reduced when appropriate releasing agents are applied. Additionally, we found that there is a relationship between the C-foil thicknesses and contact angles of pure water droplets on the substrate treated with an adequate releasing agent. ILs are potentially promising candidates for application as releasing agents when used in conjunction with an appropriate method.

References

1. D.N. Braski, Nucl. Instr. and Meth. 102(1972) 553.
2. H. Hasebe et al., Proceedings of 17th International Conference on Cyclotrons and Their Applications (Cyclotrons 2004), Tokyo, Oct. 2004. p.313.
3. H. Hasebe et al., Nucl. Instr. and Meth. **A 655** (2011) 57.
4. DAIKIN INDUSTRIES, Ltd., URL: https://www.daikinchemicals.com/solutions/products/mold-release-agents.html
5. Shin-Etsu Chemical Co., Ltd., URL: http://www.shinetsusilicone-global.com/products/type/mold/index.shtml
6. Provided by KOEI Chemical Co., Ltd., URL: https://www.koeichem.com/en.html

Preparation of different carbon stripper foils and application in beam diagnostics

Bettina Lommel[1*], *Elif* Celik Ayik[1], *Annett* Hübner[1], *Birgit* Kindler[1], *Thomas* Reichert[1], *Rahul* Singh[1], *Jutta* Steiner[1], *Beata* Walasek-Höhne[1], and *Vera* Yakusheva[1]

[1]GSI Helmholtzzentrum für Schwerionenforschung mbH, 62391 Darmstadt, Germany

Abstract. The production of carbon stripper foils that are mounted at different locations of the heavy-ion accelerator is an important task of the target laboratory, and the process for the preparation of carbon foils is constantly improved. Recently, we tested additional heat treatment of the carbon stripping foils following our standard procedure to reduce stress in the foils, since the performance and the durability of these stripping foils is a crucial factor for effective operation. We investigated the properties of foils produced some time ago and of foils freshly produced with the standard procedure, and foils treated thermally after the standard procedure. For the beam diagnostic department at GSI radiation resistant target materials for transverse ion beam profile measurement is an ongoing research topic. In that framework, the feasibility of Optical Transition Radiation (OTR) generation due to ion beam interaction with metal target for beam profile determination has been investigated. Here, carbon stripper foils are applied to enhance the OTR signal. The differently prepared foils were applied in many different experiments of the beam diagnostic department. We describe the performance of the foils during irradiation and compare the features of foils before and after irradiation.

1 Introduction

In the target laboratory of GSI, we produce carbon foils by resistance heating for the last two decades [1]. We advance the process constantly; especially refining the thermal treatment of the foils at several stages during the whole process to reduce the stress in the carbon layers and enable the production of thicker carbon foils [2].

An important application for carbon foils is as stripper foils needed at different locations of the GSI accelerator to increase the charge state of the ions for further acceleration. The strain on the foils introduced by the heavy-ion beam can be high and the durability of the stripper foils is a major challenge [3]. When the beam travels from the linear accelerator to the transfer channel towards the synchrotron, it passes through the foil stripper, where carbon foils from 200 μg/cm² up to 1000 μg/cm² are mounted. The highest requirement for the stripper foils is the charge stripping of the uranium beam from U^{+28} to U^{+73}.

We developed an additional heat treatment of the separated foils and compared the performance of these foils as stripper foils with those produced following our standard procedure. The enhanced durability of the stripper foils produced with the additional heat treatment opens the opportunity for a new application in beam diagnostics to serve as a beam monitor in parallel to the stripping function.

2 Production of carbon stripper foils

As described in detail in Kindler et al. [2], we can produce carbon foils by resistance heating with a thickness of up to 600 μg/cm² or even thicker today. In one evaporation run, we can add about 100 μg/cm² thickness to the carbon layer with each carbon rod. For releasing thick carbon foils successfully, we developed a process of cooling, relaxation and thermal treatment: The coated glass plates remain in the evacuated chamber for 24 h to allow the layers to relax. Then we dismount the plates, place them in a furnace for 1 h at 100 °C in air and allow them to cool down to room temperature afterwards. In a second step, we heat the coated plates for 4 h at 130 °C and allow them to cool down and relax overnight. Now, the foils can either be floated directly very slowly or they are stored - still on the glass plates - in a drying cabinet until needed.

Because of the hygroscopic nature of the release agent betaine sucrose, it is crucial for the handling and thermal processing of the coated glass plates that our laboratory has a ventilation that provides for stable temperature and stable relative humidity of about 32 %.

Nevertheless, the carbon foils produced by this process, now standard procedure, still have intrinsic strains, as can be seen in figure 1, visible by the curling of the foils. We can glue these foils flat on stripper frames without problems; an example is shown in figure 2. For this, we uncurl the foil onto the frame and apply carbon glue (Leit C nach Glocke from PLANO, with

* Corresponding author: b.lommel@gsi.de

https://doi.org/10.1051/epjconf/202328505002

butyl acetate as solvent) with a fine brush on the edges of the foil to fix the foil on the frame.

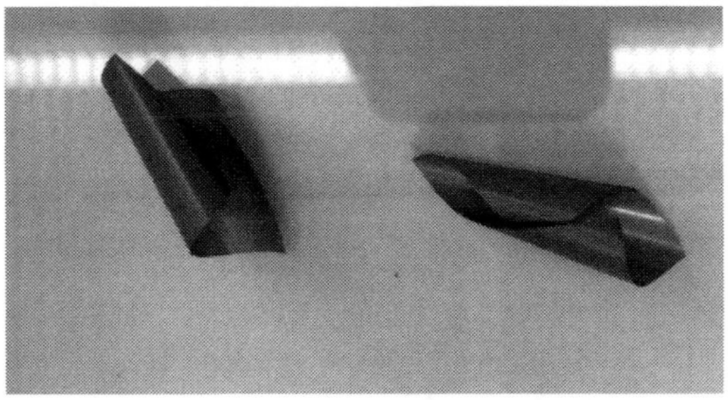

Fig. 1. Carbon foils 570 µg/cm² thickness released from glass plate, prepared by standard procedure.

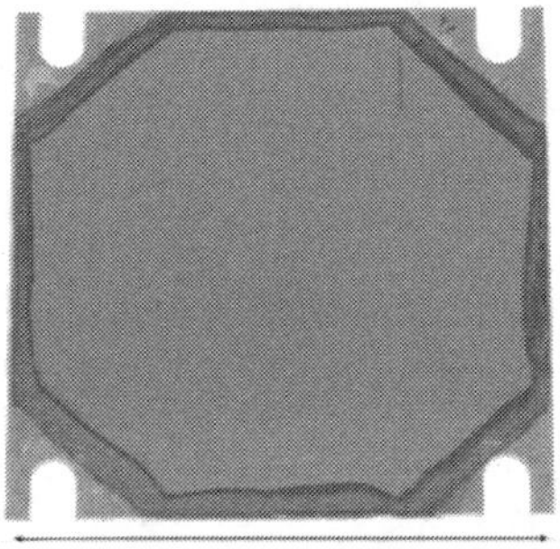

Fig. 2. Carbon foil with 570 µg/cm² thickness on a stripper frame.

Charge stripping with heavy ion beams is an ongoing research topic for several decades now, and currently carbon foils produced by our department with beam spot scanning provides the longest and most reliable operation in terms of stripped beam transmission Nevertheless, the stripper foils deform significantly during irradiation, as clearly demonstrated in figure 3. You see three of those stripper foils on a ladder inside the beam diagnostic chamber. The two foils on the left have been irradiated with uranium, whereas the foil on the right side was not irradiated. In the centre, where the beam deposited most of its intensity, the foils bulge strongly.

Fig. 3. Carbon foils with 570 µg/cm² thickness in the beam diagnostic chamber. The two foils on the left have been irradiated with U^{28+}.

To release more stress from the carbon foils, we tested an additional heat treatment of the already released foils. For this procedure, we carefully uncurled the rolled-up foils and held them flat by weighing the edges down with glass slides, as depicted in figure 4 on the left side. We then annealed these foils at 180 °C for 1 h in air. As a result, the carbon foils lie nearly flat after the 2nd annealing step, as shown in figure 4 on the right side.

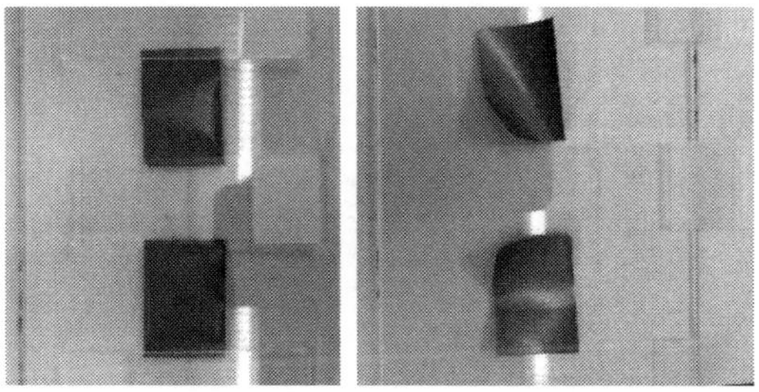

Fig. 4. Carbon foils 570 µg/cm² - Left: Stretched with glass slides before 2nd annealing step. Right: Stress-relieved foils after annealing.

Those foils were tested as strippers at the accelerator. They performed well without showing significant changes in surface structure or form, as can be seen in figure 5. The crack in the foil in the middle happened during removing of the ladder from the beam diagnostic chamber and was not caused by irradiation with the beam.

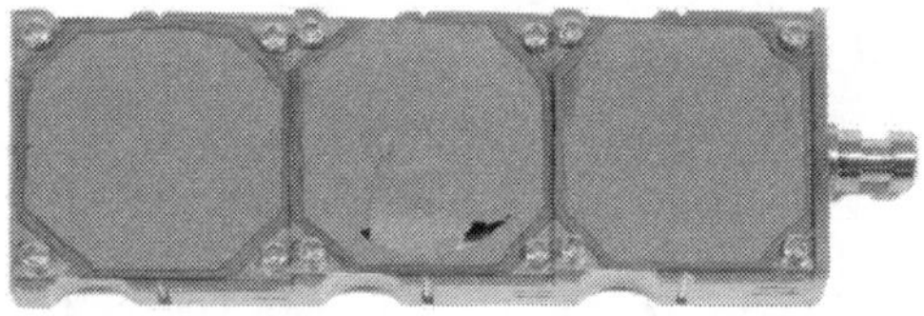

Fig. 5 Carbon foils with 570 µg/cm² thickness and treated with 2nd annealing process in the beam diagnostic chamber after irradiation with U^{28+}

For the irradiation tests, the beam parameters were comparable for each foil. The beam was delivered by GSI UNILAC operating with 36 MHz pulsed RF system. The beam macro-pulses were 300 µs long with a repetition rate of 1 Hz. Average macro-pulse had an integrated charge of ~ 1.5 nC ($3*10^8$ U^{28+} ions) with kinetic energy of 11.4 MeV/u. The uncertainty in the beam energy was below 0.1 %. The beam current and thus the integrated charge exposure variation under high current uranium operation was below 10 % for the integrated time in our experiment.

A notable observation was that the deformation of non-thermally treated foils occurred immediately after few heavy ion pulse shots on the foil and remained stable for the rest of the beam time. For thermally treated foils, deformation was barely noticeable. We operated all these foils for approximately 4 hours each during these experimental campaigns for beam diagnostics,

This improved shape retention during irradiation leads to a significant improvement in durability of the stripper foils. In addition, this flat surface opens the possibility to apply the stripper foils as beam monitoring device as well.

3 Optical transition radiation OTR

Parallel to the charge stripping of the beam, the beam diagnostics department operates an experimental setup to develop and test new methods for the detection of the beam.

When an ion crosses the interface of two media of different electromagnetic properties, Optical Transition Radiation (OTR) is produced. As described in equation (1), the number of emitted photons I is dependent on the charge state of the ion q, the velocity of the ion β and of the number of ions N crossing the interface.

$$I \sim q^2 \cdot \beta^2 \cdot \mathrm{N} \qquad (1)$$

Therefore, OTR signals can be applied in beam diagnostics for monitoring the beam properties like e.g. beam position or beam profile. The beam diagnostics department applied this concept at GSI already [4, 5].

In figure 6, the OTR signal for an uranium beam with and without a charge stripping foil is shown, visible on the OTR monitor target.

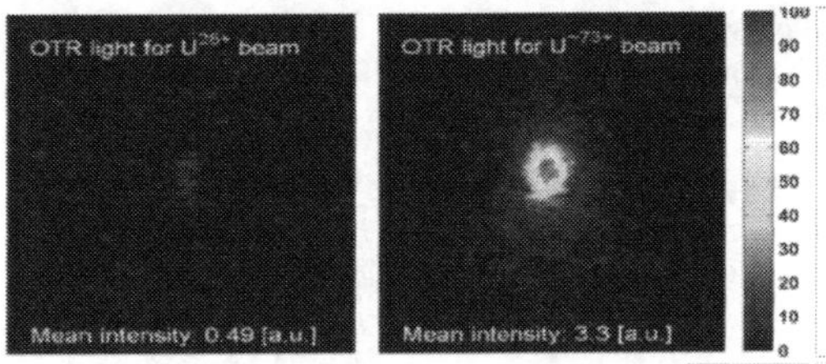

Fig. 6. OTR signal without and with carbon stripping foil.

In principle, an OTR signal is visible on the stripper foil as well as on the beam monitor. Nevertheless, for the evaluation of the OTR signal as a measure of the charge distribution in the beam, an undistorted projection in two dimensions is needed. Carbon foils produced with the described production process, stay stable and flat during irradiation, so the visible two-dimensional charge distribution can be evaluated.

4 Outlook

For the future upgrade, it is foreseen to apply the carbon foil as stripper and as beam profile monitor simultaneously at UNILAC intensities. With this dual usage possibility of the foils, the total matter in the ion beam path will be reduced. This is a helpful feature during the accelerator operation in the long-term science user experiments also referred to as "production mode" both from the perspective of accelerator maintenance and robustness of beam parameters achieved.

Consequently, the charge distribution of the beam can be observed permanently during production without additional beam diagnostic monitors.

References

1. Preparation of self-supporting carbon thin films; B. Lommel, W. Hartmann, B. Kindler, J. Klemm and J. Steiner; Nucl. Instr. and Meth. A 480, 199-203 (2002).
2. Development of carbon foils with a thickness of up to 600 µg/cm^2, Birgit Kindler, Willi Hartmann, Annett Hübner, Bettina Lommel, Jutta Steiner; Nucl. Instr. and Meth. A 613, 126-130 (2010).
3. W. Barth, M. Kaiser, B. Lommel, M. Maier, S. Mickat, B. Schlitt, J. Steiner, M. Tomut, H. Vormann, Journ Rad. Nuc. Chem 299 (2) 1047-1053 (2014).
4. Beata Walasek-Höhne et al. „Pilot Studies on Optical Transition Radiation Imaging of Non-Relativistic Ions at GSI" BIW 2012 Proceedings.
5. R. Singh and T. Reichert; "Longitudinal charge distribution measurement of nonrelativistic ion beams using coherent transition radiation". Phys. Rev. Accel. Beams 25, 032801 – Published 23 March 2022.
6. R. Singh, T. Reichert and B. Walasek-Hoehne; „ Optical transition radiation based transverse beam diagnostics for nonrelativistic ion beams". Phys. Rev. Accel. Beams 25, 072801 (2022) DOI:https://doi.org/10.1103/PhysRevAccelBeams.25.072801

EPJ Web of Conferences **285**, 05003 (2023) https://doi.org/10.1051/epjconf/202328505003
INTDS2022

Installation and commissioning of the stripping foil system in the new CERN PS booster 160 MeV H⁻ injection region

Wim Weterings[1*], *Chiara* Bracco[1], *Louise* Jorat[1], and *Elisabeth* Renner[2]

[1]CERN, 1211 Geneva 23, Switzerland
[2]TU Wien, Vienna, Austria

Abstract. During Long Shutdown 2 (LS2) at CERN, the new Linac4 (L4) accelerator was connected to the PS Booster (PSB) to inject 160 MeV H^- beam into the 4 superposed PSB rings. In order to achieve this, a completely new H^- charge-exchange injection chicane system, with 200 μg/cm² carbon stripping foil units, to convert the negative hydrogen ions into protons, by stripping off the electrons, has been installed. Beam commissioning of this system started in December 2020 and different types of foils have been used for this purpose. In parallel, stripping efficiency measurements, of different foil types, continued to take place with the stripping foil test stand in the L4 Transfer Line. This paper briefly recalls the final design, and installation of, the main components in the PSB injection region, before reporting on the important results obtained during measurements and commissioning with beam.

1 Introduction

The CERN Proton Synchrotron Booster (PSB) is the first synchrotron in the CERN injector complex and is used for tailoring the wide range of transverse beam characteristics requested by the various users at CERN. To produce the challenging High Luminosity Large Hadron Collider (HL-LHC) beam parameters, a massive improvement program of the injector chain was completed in 2021, under the LHC Injectors Upgrade (LIU) project [1]. The project included a new Linac, so-called Linac4 (L4) [2], as well as major upgrades and consolidation of the Proton Synchrotron Booster (PSB), the Proton Synchrotron (PS) and the Super Proton Synchrotron (SPS).

L4 is a linear accelerator delivering negative hydrogen ions (H^-) to the PSB at a higher injection energy of 160 MeV than the previous 50 MeV protons (H^+) beam of Linac2. Therefore, the PSB was upgraded in 2019/2020 during the Long Shutdown 2 (LS2), a major aspect of this upgrade was the connection of the L4 to the PSB, with an H^- charge-exchange injection system [3], replacing the conventional H^+ multi-turn injection.

Commissioning of the upgraded PSB started in December 2020 and was completed in Q1 2021 [4].

2 PSB injection modifications

The PSB consists of 4 superposed synchrotron rings and, in the new configuration, the 160 MeV H^- beam from L4 is distributed to the four rings by a vertical bending magnet (DVT30), a system of 5 kicker magnets (DIS), a vertical bend (DVT40) and 3 septum magnets (SMV) [5]. The beam is subsequently injected horizontally into the PSB by a H^- charge-exchange injection system, one for each ring.

Charge-exchange injection can achieve higher particle density providing an extremely flexible way to load particles into the PSB, which has a revolution period $\tau \approx 1\,\mu s$ at 160 MeV, making the accumulation of many injected turns possible with a tight control of the beam density [6]. In the new injection system, H^- are progressively injected horizontally into the PSB and converted into H^+ by passing through a 200 μg/cm² carbon foil to strip off the electrons, aiming to convert at least 98% of the beam to H^+ [7]. Partially stripped H^0 and ≤ 1% H^- missing the foil are directed to internal Ti_6Al_4V dumps, one per ring, equipped with H^0/H^- monitors [8] to measure the stripping efficiency and interlock the injection on any foil degradation or failure.

During injection, the local orbit of the PSB circulating beam is displaced 46 mm horizontally by a set of 4 pulsed dipole magnets (BSW) [9] to merge with the injected beam (Fig. 1).

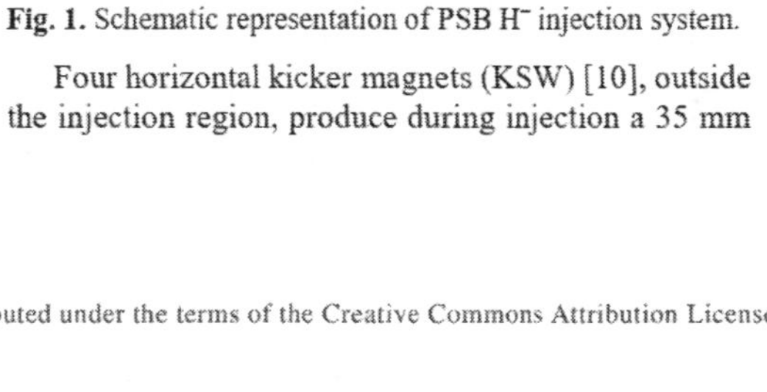

Fig. 1. Schematic representation of PSB H⁻ injection system.

Four horizontal kicker magnets (KSW) [10], outside the injection region, produce during injection a 35 mm

* Corresponding author: wim.weterings@cern.ch

closed orbit bump, with falling amplitude, to paint the beam into the required horizontal emittance and move the circulating beam away from the stripping foil.

3 Foil exchange mechanism

The stripping foil handling and exchange mechanism, presented in detail in [11, 12], consists of a stainless-steel belt, rotating over two pulleys, to which a maximum of six foil holders can be attached by use of quick disconnect sliders (Fig. 2). This allows moving foils into the beam aperture, with a perpetual rotation; each of the six foils can be re-selected into the nominal beam position with a possible horizontal foil adjustment of ±2 mm, and a precision of ±0.1 mm, to find the optimum position, while having precise measurement of the foil position of over this 4 mm range [13].

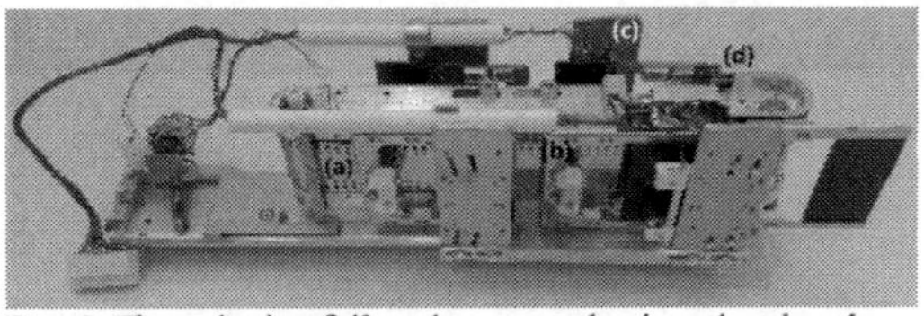

Fig. 2. The stripping foil exchange mechanism showing the rotating stainless-steel belt (a), the holders with stripping foils attached (b), UHV compatible microswitches (c) and membrane potentiometers (d) for precise positioning.

4 PSB Beam commissioning

Beam commissioning of the upgraded PSB and new H^- charge-exchange injection region started in December 2020, including foil efficiency measurements and optimisation of the beam production schemes for tailoring the various user-defined beam requirements.

4.1 Stripping efficiency measurement

In 2018/2019 foils of different manufactures have been qualified regarding gluing & manipulation and stripping efficiency & lifetime in a test stand [11] installed in the L4 transfer line [14, 15], described in more detail in § 6. The selected stripping foils for PSB beam commissioning are shown in Table 1.

Table 1. Characteristics of foil used for PSB commissioning.

Foil #	Description	Thickness	Reference	
1, 4	Amorphous Carbon*	200 μg/cm^2	XCF-200	[16]
2, 5	Amorphous Carbon*	200 μg/cm^2	GSI-200	[17]
3, 6	Multilayer Graphene	240 μg/cm^2	MLG-250	[18]

* Arc evaporated amorphous Carbon

By comparing the Beam Current Transformers (BCT) positioned in the L4 transfer line and the PSB ring, the beam transmission can be calculated, which is dependent on the foil efficiency as well as other loss mechanism such as resonance crossing, instabilities, aperture bottlenecks, etc. Figure 3 shows the average BCT signal, for each PSB ring, considering all beam types, with a 98% target below which alerts, or beam interlock might take place. Beam Loss Monitors (BLM) values, localised in the injection region, are also shown.

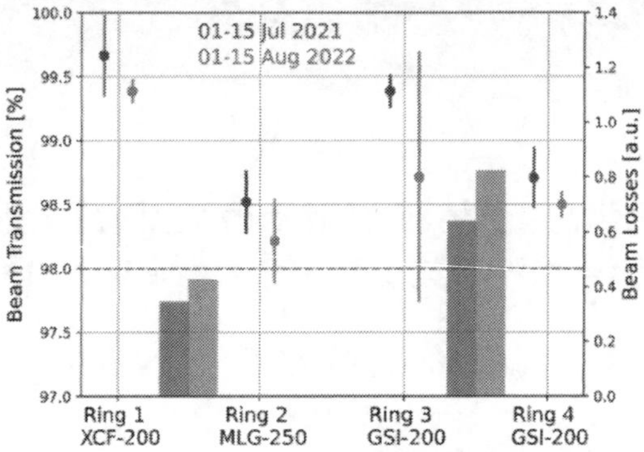

Fig 3. Beam transmission (dots) and losses (bars) at injection, measured with BCT positioned in the L4 transfer line and the PSB ring, as well as BLM in the injection region.

The degradation and large standard deviation of the beam transmission at Ring 3 was analysed in detail by looking at the signals from the H^0/H^- monitor [8] for different PSB beam types. All foils showed a stripping efficiency of at least 99.5%, but large signal fluctuations were measured for the different beams at the H^- plates while H^0 remained constant (Fig. 4). In general, one would expect a higher signal of H^0 than H^-, but since this was not the case, there was a clear indication that the stripping efficiency was influenced by the steering of the beam, thus H^- ions not intercepting the foil. The stripping efficiency could indeed be brought back to the original ~99.5% value by adjusting the trajectory of all beams at the foil.

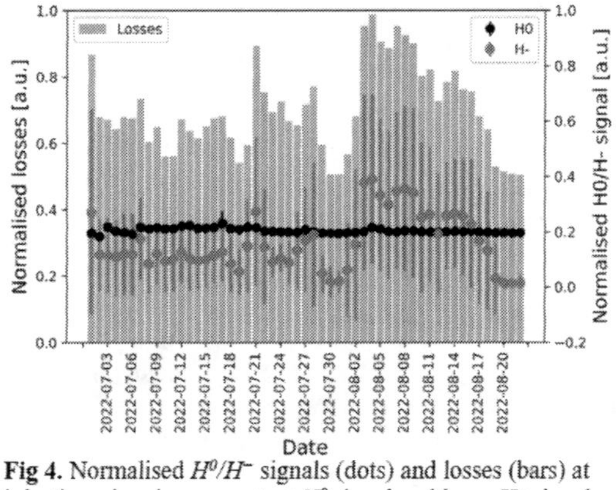

Fig 4. Normalised H^0/H^- signals (dots) and losses (bars) at injection, showing a constant H^0 signal and large H^- signal fluctuations until 19/08/2022, after which proper steering of the beam remedied this issue.

4.2 Emittance measurement

The choice of 200 μg/cm^2 stripping foil is based on providing sufficient stripping efficiency, while minimizing the emittance blow-up, and hence the degradation of the beam brightness, induced by foil scattering [6]. The foils installed in the PSB were qualified by single foil passages in the L4 test stand [11], verification of the predicted scattering properties was thus not possible.

During PSB commissioning, the flexibility to customize the number of foil crossings enabled the assessment of the emittance increase induced by multiple passages through the foils [4]. Figure 5 (left) illustrates the emittance increase, measured for a varying number of foil crossings (N_t) of the GSI-200 foil. Analytical and simulation results show good agreement with the measurements for $N_t > 30 - 50$. Figure 5 (right) also shows the measurements for $N_t = 150$ for the different foil types as listed in Table 1.

For all tested foils, the measured emittance increase is consistent with the model. No significant foil induced beam degradation is expected for any foil type to produce high brightness beams (10 to 35 injected turns).

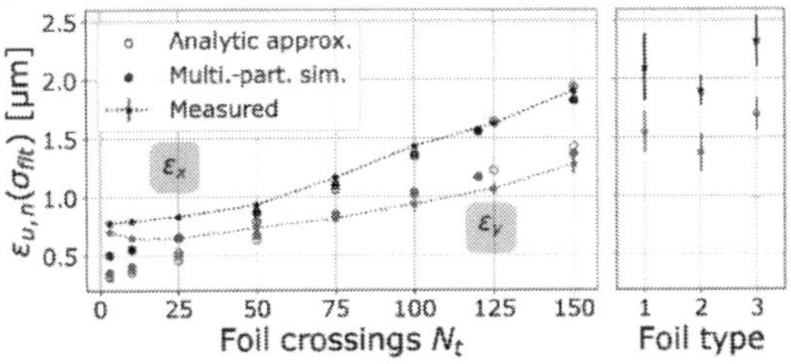

Fig 5. Left: Transverse emittances for varying number of foil hits N_t for foil #2 GSI-200. Right: Different foil types, as described in Table 1 for $N_t = 150$ [4].

5 Observations

In [14], we described the sudden GSI-200 foil deformation, towards the centre of beam impact, after only one pulse in the L4 test stand. The same phenomenon was also observed during PSB beam commissioning, but only at higher injected intensities, and dedicated beam time was used to investigate.

While increasing in small steps the injected intensity, whilst recording with the BTV camera [11, 12], increasingly small local deformation was observed at the beam spot. Reaching up to 70 injected turns, a sudden light wave was seen, during which the foil deformation took place, see Fig. 6, but this had no influence on the stripping efficiency.

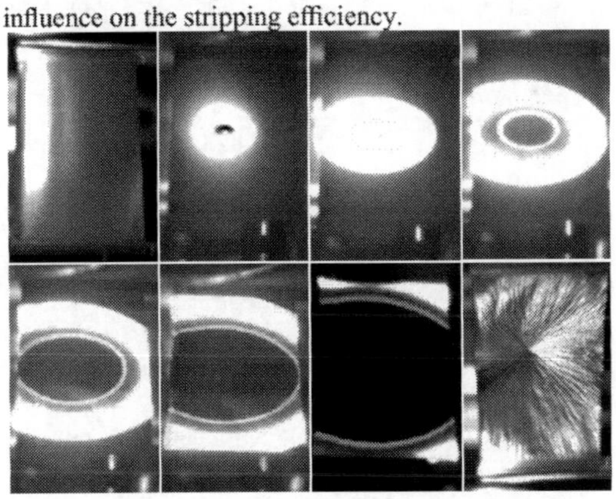

Fig. 6. GSI-200 foil before (top-left) and after (bottom-right) beam impact with images of light wave observed on foil surface at ms intervals (left to right). Event took place with 330 ns, 25 mA beam pulses, 70 injected turns with 5E10 H^- per turn and 150 foil crossings (~6.7E12 H^- foil passages).

One hypothesis is that residue of the betaine-sucrose parting agent, left on the foil surface, ignites and burns off creating this effect [19], but other experts [20] mentioned that this is observed on other foils and thought to be linked to grain friction reaching the carbon diffusion temperature.

6 L4 test stand

A stripping foil test stand [11] is installed in the L4 transfer line allowing measurement, at the % level, of the stripping efficiency for different foils by measuring the decrease in current between two cross-calibrated BCTs installed upstream and downstream of the test stand. To qualify suppliers for possible future orders, measurements were done during weeks 38 & 48 2020.

The beam characteristics during the tests are shown in Table 2, and the foils used are shown in Table 3.

Table 2. Characteristics of the beam used for the tests.

Parameter	Value	Unit
Ion Species	H^-	
Energy	160	MeV
Current	25	mA
Pulse Length	500*	μs
Repetition	1.2	s

*4 x 125 μs batches, 1 μs gap

Table 3. Characteristics of foils used in L4 test stand.

Foil #	Description	Thickness	Reference	
1	Amorphous Carbon*	200 μg/cm²	XCF-200	[16]
2	Amorphous Carbon*	200 μg/cm²	GSI-200	[17]
3	Multilayer Graphene	200 μg/cm²	NAN-200	[21]
4	Graphene	200 μg/cm²	AGP-200	[22]
5	Graphene	200 μg/cm²	SHT-200	[23]
6	Graphene rGO	185 μg/cm²	ARC-200	[24]

* Arc evaporated amorphous Carbon

The transmission data of the tests are shown in Fig. 7. In comparison to reference foils #1 & #2, also installed in the PSB injection, #3 & #6 have a very good transmission with a comparable 99% efficiency. The performance of #5 is slightly less, about 98%, and #4 only has a 94.5% efficiency. Improved samples of #4 have been received for further testing and #6 was identified as a possible future (European) supplier.

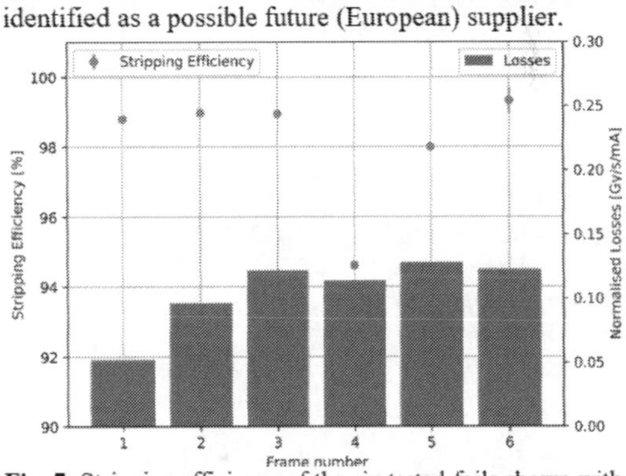

Fig. 7. Stripping efficiency of the six tested foils shown with the corresponding normalised beam losses.

7 Conclusion

A completely new H^- charge exchange injection system comprising of chicane magnets, painting bump kickers, stripping foil exchange mechanisms, internal beam dumps and dedicated instrumentation has been installed, and successfully commissioned, in the 4 superposed PSB rings. The installed stripping foils were qualified using the L4 test stand and all show the expected theoretical efficiency > 98%. Measurements of emittance increase, caused by multiple beam passages through the foils, show that no significant foil induced beam degradation is expected to produce high brightness beams.

Observed sudden visual degradation of foils was studied in detail and is thought to be linked to sudden combustion of residue left on the foil or inter-grain friction reaching the carbon diffusion temperature.

Furthermore, additional stripping efficiency tests with the L4 test stand yielded in encouraging results to qualify possible new suppliers.

The authors would like to thank everybody involved in the successful installation and commissioning of the PSB 160 MeV H^- injection region for their valuable contribution, but in particular R. Noulibos for his continuous engineering efforts and ingenious ideas, M. Plum for initiating us to the world of stripping foils, the INTDS members for sharing their valuable expertise and knowledge and the CERN Control Centre (CCC) operators for giving support and input during hardware and beam commissioning.

References

1. M. Meddahi et al., *LHC Injectors Upgrade Project: Towards New Territory Beam Parameters*, in Proceedings IPAC'19, Melbourne, Australia, May 2019, pp. 3385-3390 (2019) doi:10.18429/JACoW-IPAC2019-THXPLM1
2. A.M. Lombardi, *Linac4: From Initial Design to Final Commissioning*, in Proceedings IPAC'17, Copenhagen, Denmark, May 2017, pp. 1217-1222 (2017) doi:10.18429/JACoW-IPAC2017-TUYA1
3. W. Weterings, C. Bracco, L. Jorat, M. Meddahi, R. Noulibos, P. Van Trappen, *The New Injection Region of the CERN PS Booster*, in Proceedings IPAC'19, Melbourne, Australia, May 2019, pp. 2414-2417 (2019) doi:10.18429/JACoW-IPAC2019-WEPMP039
4. E. Renner et al., *Beam Commissioning of the New 160 MeV H^- Injection System of the CERN PS Booster*, in Proceedings IPAC'21, Campinas, SP, Brazil, May 2021, pp. 3116-3119 (2021) doi:10.18429/JACoW-IPAC2021-WEPAB210
5. W. Weterings et al., *Status of the 160 MeV H− Injection into the CERN PSB*, in Proceedings IPAC'12, New Orleans, USA, May 2012, TUPPR091, pp. 2041-2043 (2012)
6. J.L. Abelleira et al., *Painting Schemes for CERN PS Booster H^- Injection*, in Proceedings IPAC'15, Richmond, USA, May 2015, pp. 3879-3882 (2015) doi:10.18429/JACoW-IPAC2015-THPF083
7. B. Goddard, M. Aiba, C. Bracco, C. Carli, M. Meddahi, W. Weterings, *Stripping Foil Issues for H^- Injection into the CERN PSB at 160 MeV*, in Proceedings IPAC'10, Kyoto, Japan, May 2010, THPEB030, pp. 3951-3953 (2010)
8. A. Navarro Fernandezet al., *PSB H^0-H^- Monitor Calibration and Commissioning*, in Proceedings IBIC'21, Pohang, Rep. of Korea, May 2021, pp. 429-433 (2010) doi:10.18429/JACoW-IBIC2021-WEPP24
9. B. Balhan, C. Baud, J. Borburgh, M. Hourican, *Design and Construction of the CERN PS Booster Charge Exchange Injection Chicane Bumpers*, in Proceedings IPAC'18, Vancouver, Canada, April 2018, pp. 2575-2577 (2018) doi:10.18429/JACoW-IPAC2018-WEPMF082
10. L.M.C. Feliciano et al., *A New Hardware Design for PSB Kicker Magnets (KSW) for the 35 mm Transverse Painting in the Horizontal Plane*, in Proceedings IPAC'15, Richmond, VA, USA, May 2015, pp. 3890-3892 (2015) doi:10.18429/JACoW-IPAC2015-THPF086
11. W. Weterings, C. Bracco, R. Noulibos, Y. Sillanoli, P. van Trappen, J. Radioanal. Nucl. Chem. **305**, pp. 831-835 (2015) https://doi.org/10.1007/s10967-014-3917-0
12. W. Weterings, C. Bracco, L. Jorat, R. Noulibos, P. van Trappen, AIP Conference Proceedings **1962**, 030003 (2018) https://doi.org/10.1063/1.5035520
13. P. Van Trappen, R. Noulibos, W. Weterings, *Stripping Foil Displacement Unit Control for H^- Injection in PSB at CERN*, in Proceedings ICALEPCS'15, Melbourne, Australia, October 2015, pp. 363-366 (2015) doi:10.18429/JACoW-ICALEPCS2015-MOPGF121
14. W. Weterings, C. Bracco, L. Jorat, R. Noulibos, P. van Trappen, EPJ Web Conf. **229** (2020) 01003, https://doi.org/10.1051/epjconf/202022901003
15. C. Bracco et al., *Commissioning of the Stripping Foil Units for the Upgrade of the PSB H^- Injection System*, in Proceedings IPAC'17, Copenhagen, Denmark, May 2017, pp. 595-598 (2017) doi:10.18429/JACoW-IPAC2017-MOPIK041
16. ACF-Metals, 2239 E. Kleindale Road, Tucson, Arizona, U.S.A.
17. GSI GmbH, Planckstraße 1, 64291, Darmstadt, Germany.
18. KANEKA Corporation, 5-1-1, Torikai-Nishi, Settsu Osaka 566-0072, Japan.
19. Video conference with B. Lommel, GSI GmbH, Darmstadt, Germany, 12th May 2022.
20. Discussion during INTDS2022 conference with V. Jaggi, Micromatter Technologies inc., Surrey, Canada, 27th September 2022.
21. Applied Nanotech, Inc., 3006 Longhorn Blvd., Suite 107, Austin, USA.
22. Advanced Graphene Products Sp. z o.o., 66-002 Zielona Góra, Poland.
23. SHT Smart High Tech AB, Kemivägen 6, 412 58 Gothenburg, Sweden.
24. L. Megido-Fernandez, C. Botas and C. Blanco.

https://doi.org/10.1051/epjconf/202328506001

Characterization of a Nb-Pb composite target

Anand Pandey[1, *], *Ravi* Bhushan[1], *Megha* Pant[1], *Ajit* Mahapatro[1], *Aman* Rohilla[2], *Chandrani* Majumder[3], *Rajesh Pratap* Singh[4], *Sunil* Ojha[4], *Ganjigatte* Umapathy[4], *Leema* Saikia[4], and *Sanjay Kumar* Chamoli[1]

[1]Department of Physics and Astrophysics, University of Delhi, New Delhi-110007, India
[2]Institute of Space Sciences, Shandong University, Weihai 264209, People's Republic of China
[3]Department of Physics, Indian Institute of Technology Bombay, Mumbai-400076, India
[4]Inter University Accelerator Center, Aruna Asaf Ali Marg, New Delhi-110067, India

Abstract. A thin Nb target on a Pb backing was fabricated using the rolling technique. In the present work, the characterization of the sample has been carried out using various techniques to study the surface morphology, crystallography, elemental analysis, thickness and electrical conductivity useful for the intended use of the target prepared. The thin Nb target was analyzed using scanning electron microscopy (SEM), energy dispersive X-ray spectroscopy (EDS), X-ray diffraction (XRD) and Rutherford back-scattering spectroscopy (RBS) techniques. The uniform and smooth surface of the sample was confirmed by SEM imaging. The presence of oxygen in the EDS spectra and peaks corresponding to crystal planes of PbO_2 in the XRD spectrum confirms the oxidation of Pb foil. The XRD spectrum shows clear peaks corresponding to (011), (002), (112) crystal planes of pure Nb. The thickness of the thin Nb foil measured using the RBS techniques was found to be 0.8 mg/cm^2. The current(I)-voltage(V) curve (relationship between the current flowing through an electronic device and the applied voltage across its terminals) at room temperature for the Pb and Nb/Pb foils was plotted and showed linear relationship between applied voltage and current flowing through the target foil. The measured electrical conductivity indicates an enhancement of 35.9 % for Nb/Pb foil relative to the conductivity measured for the Pb foil.

1 Introduction

Targets of specific thicknesses and physical properties are crucial for nuclear experiments. Lifetime measurements require different types of targets depending on the lifetime of the states (nanoseconds to sub-picosecond regime) to be measured and the respective technique involved in the measurement. The recoil distance doppler shift (RDDS) method [1] ideally requires thin self-supporting targets (without backing) so that populated nuclei recoils in the vacuum while the doppler shift attenuation method (DSAM) [2] requires thin target with the relatively thick backing of high-Z material so that populating nuclei recoil in the backing medium. The nuclear physics experiment depends heavily on the quality of the target, especially for DSAM lifetime technique; there must not be any vacuum gap between the target and backing foil [3]. In the experiment, the target is bombarded with high energy heavy ion beams and enormous heat is produced on the surface of the target, which needs to be ejected out to the surroundings in time to prevent any possible damage of the foil during the experiment. In view of the complexity of such an experiment, highly pure target foils of appropriate thickness having high melting point as well as high electrical and thermal conductivity is of great importance.

In the reported work, a thin Nb target with a thick Pb backing was fabricated using the mechanical rolling technique [4] for lifetime measurements to extract lifetimes of excited nuclear states in ^{118}Xe with the DSAM technique. The main objective of this work is to determine the quality of the fabricated target from the viewpoint of its intended use. Purity, elemental analysis, crystallinity, surface morphology and thickness are monitored using different material characterization techniques while electrical conductivity was determined using the I-V curve.

2 Fabrication of the target

A Nb foil of ~ 1 mg/cm^2 thickness was prepared using mechanical rolling. The Pb foil of thickness ~ 13 mg/cm^2 was prepared using the one-way rolling method as Pb is extremely soft in nature and two directional rolling often damages the foil. Finally, both the foils, Nb and Pb, are kept over one another placed inside the stainless steel (SS) pack for further rolling to form the composite target. Detailed experimental procedure of fabrication has been reported previously [3].

* Corresponding author: pandeyanand799@gmail.com

3 Characterization Techniques

3.1 Scanning Electron Microscopy (SEM)

The surface morphology of the target foil was imaged using the field emission scanning electron microscopy (FESEM, MIRA3, Tescan, USA) operated at 30 kV, present at University Science Instrumentation Centre (USIC), Department of Physics and Astrophysics, University of Delhi, Delhi. SEM provides the images by raster scanning the surface of the sample using focused beam of electrons. The images produced using electron microscopy are shown in Figure 1(a) and Figure 1(b). The SEM image (Figure 1(a)) suggests the uniform and smooth surface of the foil obtained in the present work while Figure 1(b) shows the magnified part of the foil where two different regions, grey (smooth Nb surface) and bright (granular Pb) were observed. This was also confirmed by the elemental composition study using the EDS technique.

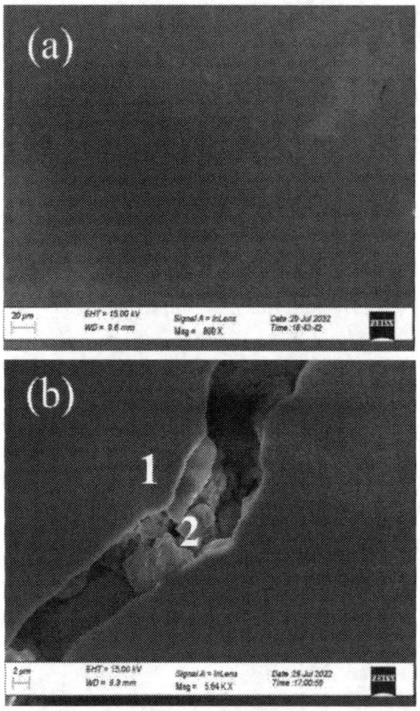

Fig. 1. FESEM images of the fabricated target foil: (a) Nb surface with x800 zoom, (b) Nb surface with x5000 zoom, showing marked regions for elemental composition by EDS analysis.

3.2 Energy Dispersive X-ray Spectroscopy (EDS) technique for elemental composition

The elemental composition of the target foil was studied using an EDS attachment to the FESEM system. EDS spectra taken on the two different regions marked as area 1 and area 2 in Figure 1(b) are shown in Figures 2(a) and (b), respectively.

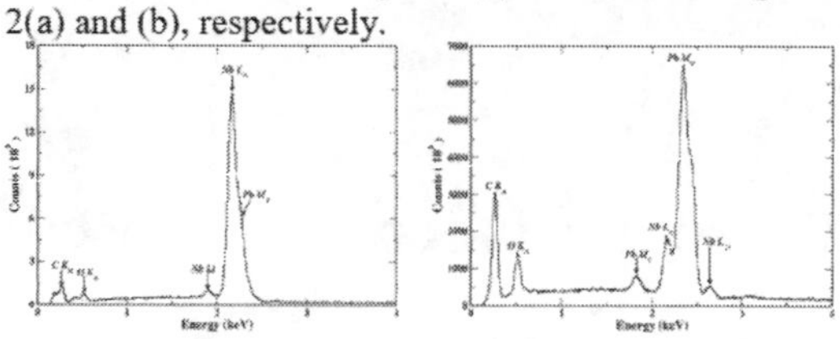

Fig. 2. EDS spectra taken at two different regions (a) Area 1 and (b) Area 2 as marked in the Figure 1(b) showing their elemental composition. All the observed peaks in the spectra are marked appropriately. EDS spectra taken at 15 kV.

The EDS spectra taken at area 1, Figure 2(a), shows a large peak of Nb(L_α), which along with the Pb(M_α) peak confirms the presence of pure Nb (no major elemental impurity), while the EDS spectra taken at area 2, Figure 2(b), shows a strong peak of Pb(M_α), which along with the peaks corresponds to niobium (Nb L_α), carbon (C K_α) and oxygen (O K_α). The observed O and C peaks may be due to oxidation or carbonization of organic contaminants on the foil surface. The results clearly indicate the uniform Nb surface over the region, because a mostly grey region was visible in the panned-out SEM images. Based on the observed EDS spectra, we can state that the fabricated target is pure (no major elemental impurity was added during the fabrication process).

3.3 X-ray Diffraction

In XRD, the incident X-ray beam interacts with the atomic layers of the sample to generate characteristic peaks corresponding to the composition, phase and internal atomic arrangement. Crystallinity and phase purity of the sample were analysed using an EMPYREAN X-ray diffractometer (manufactured by PANanalytical), present at IUAC, Delhi, with 2:1 ratio of Cu-$K_{\alpha 1}$ (λ = 1.540598 A°) and Cu-$K_{\alpha 2}$ (λ = 1.544426 A°) as X-ray source in the angular range of 2θ = 20° - 80° at a scan rate of 1°/min [5].

The resulting XRD spectra of the sample are shown in Figure 3(a). The sharp peaks at 2θ = 38.6°, 55.8°, 69.9°correspond to the (hkl) planes (011), (002) and (112), respectively, for metallic Nb, which agrees with the joint committee on powder diffraction standards (JCPDS) PDF data #98-007-6264. The peaks at 2θ = 31.3°, 52.2°, 62.1°, 65.2°correspond to the (111), (022), (113), (222) crystal planes originating from Pb backing, matching well with the JCPDS PDF data #98-005-2253. The peaks at 2θ = 44.9°, 47.8°correspond to the crystal planes (022), (220) of PbO_2, resulted due to oxidation of Pb, identical to the peaks shown in the JCPDS PDF data #98-0020362 assigned to PbO_2. The peaks at 2θ = 43.5°, 50.5°, 74.5°correspond to the crystal planes (111), (200), (220) of the Fe-Ni alloy in the ratio (3:2) [6], which is the target frame composition.

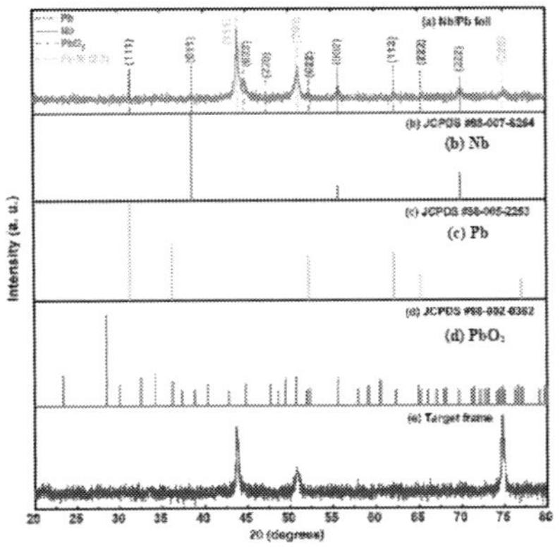

Fig. 3. XRD patterns for (a) Nb/Pb target and (e) Fe-Ni (3:2) alloy in the target frame. The line patterns representing JCPDS data card for (b) Nb (#98-007-6264), (c) Pb (#98-005-2253) and (d) PbO_2 (#98-002-0362). The peaks corresponding to the observed planes to Pb, Nb, PbO2 and Fe-Ni (3:2) alloy in the target frame are marked by the violet dash dot line, the dark cyan solid line, the red dash line and the green short dash dot line, respectively.

The estimated lattice spacing (d_{hkl}) value of 1.64 A° using Bragg's law [7] corresponding to the (002) plane for the Nb foil, agrees well with the cubic structure of Nb, as mentioned in the JCPDS PDF data #98-007-6264. The evaluated lattice constant value of 3.29 A° is consistent with the cubic structure of Nb.

3.4 Rutherford Back-scattering Spectroscopy (RBS)

To experimentally determine the thickness of thin foil, the Rutherford back-scattering technique is used. ^{4}He ions are accelerated using 1.7 MV tandem accelerator at IUAC, Delhi. When the He beam hits the target, the He ions lose energy inside the target and are back-scattered at some point inside the target due to elastic collisions. The back-scattered ion again loses energy while coming out of the target and detected by a silicon surface barrier detector (with 50 mm^2 effective area and 3.6 mSr solid angle) kept at an angle of 165° with respect to the beam direction. The detector set up is coupled with a 2k multi-channel analyzer to process the data. The analysis of RBS spectra is done by Rutherford universal manipulation program (RUMP) [8] and the simulation program for the analysis of nuclear reaction analysis (SIMNRA) [9]. The obtained energy spectrum from the RBS measurement is shown in Figure 4.

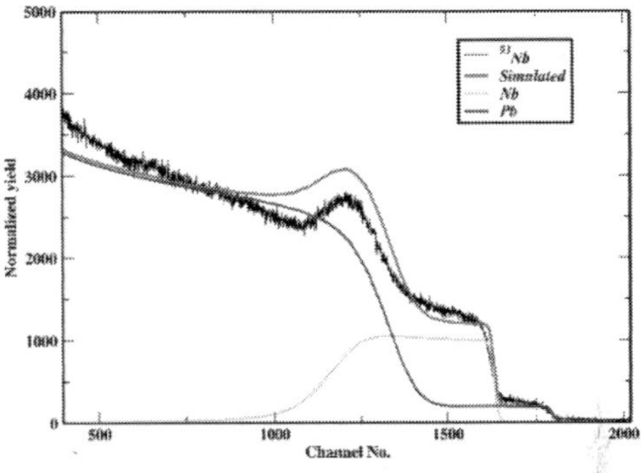

Fig. 4. RBS spectra of the Nb target with thick Pb backing fitted using XRUMP software.

The thickness of the Nb target layer obtained from the analysis is given in Table 1. The RBS spectra obtained support the results from EDS and XRD analysis, no trace of heavy elemental impurities was observed.

Table 1. Comparison of thickness of Nb foil obtained using RBS technique, energy dispersive X-ray fluorescence (EDXRF) technique and geometrical method.

S. No.	Element	RBS method (mg/cm^2)	Geometrical method* (mg/cm^2)	EDXRF [4] (mg/cm^2)
1)	Nb	0.8	1.0	0.9

(*Thickness evaluated by dividing the measured weight of foil by its approximate geometrical area.)

3.5 Electrical transport properties

As discussed in the introduction section, targets having high electrical and thermal conductivity are required to prevent any possible damage (target peeling off) during the in-beam nuclear experiment. The electrical conductivity of the sample was evaluated by measuring the current through the foil on applying voltage. Two circular conducting contacts were made on the surface of both the foils using silver paste (Ag contacts) as shown in the inset of Figure 5. Programmable Power Supply (2200-60-2 programmable power supply) was used for applying varying voltage (0 to 0.25V) across these two Ag contacts and corresponding current(I) was measured using Picoammeter (6485 Picoammeter).

The current-voltage (I-V) curves for Pb and Nb/Pb foil are shown in Figure 5(a) and Figure 5(b), respectively. The recorded I-V curve for both Pb and Nb/Pb foil shows linear dependence of current on DC voltage ($I \propto V$) [10], exhibiting ohmic conduction for charge flow for lateral device structure configurations. The resistivity was evaluated using Ohm's law as $\rho = RA/l$, where l is the distance between the Ag electrodes, A is the area of cross section, R is the resistance of the sample estimated from slope of the I-V plot shown in Figure 5. The measured values of the electrical conductivity, σ, are 172.7 S/m for Pb and 234.7 S/m for Nb/Pb foil respectively, showing an enhancement in electrical conductivity by 35.9%.

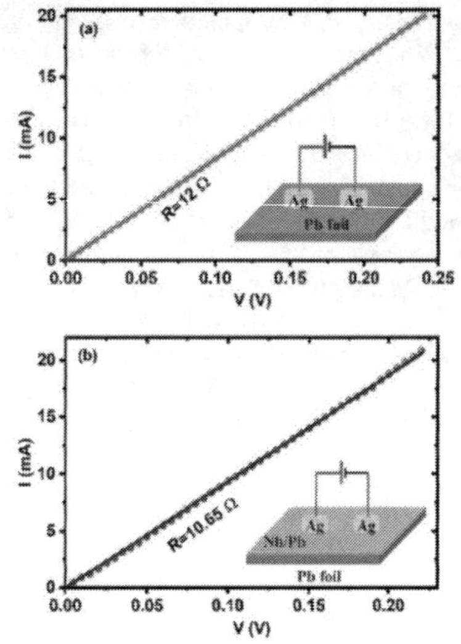

Fig. 5. I-V characteristic plots at room temperature for (a) Pb foil and (b) Nb/Pb surface connected using lateral device structure configuration using the Ag top contacts. The schematics for the measurement are shown in the respective inset.

4 Conclusion

A thin Nb target (~ 1.2 μm) on a thick Pb backing (~ 8.8 μm) was fabricated using the rolling technique. The fabricated target was successfully utilized in the nuclear lifetime measurement experiment. Characterization of the target sample was done to confirm the purity, thickness and surface uniformity. SEM images clearly indicate the surface uniformity of the target. Thickness of the target film was found out to be around 0.8 mg/cm^2 from the RBS technique. EDS, XRD and RBS analyses consistently confirm the purity of the target as no major elemental impurity was observed. Enhanced electrical conductivity by 35.9 % was measured using the I-V curve at room temperature. The results of material and electrical properties of the Nb/Pb composite foil indicate the formation of a good quality target foil that can be utilized in future envisaged nuclear experiments.

A. Pandey, acknowledges the financial support from the University Grants Commission (UGC), India. Financial support from the IUAC project (Grant No. IUAC/ XIII/7/UFR-71336) is gratefully acknowledged. Authors would like to thank IUAC for providing Target lab facility and for extending XRD facility funded by Ministry of Earth Sciences (MoES) under Geochronology project [MoES/P.O.(Seismic)8(09)-Geochron/2012]. Authors would like to acknowledge the contribution of Mr. Abhilash S.R. in preparing the target sample.

References

1. A. Rohilla et al., The European Physical Journal A **53**, 64 (2017).
2. S. Rajbanshi et al., Physics Letters B **748**, 387-391 (2015).
3. A. Pandey et al., Nuclear Instruments and Methods in Physics Research Section A: Accelerators, Spectrometers, Detectors and Associated Equipment **985**, 164667 (2021).
4. C. K. Gupta et al., AIP Conf. Proc. 1962, 030013 (2018).
5. A. Sharma et al., Nuclear Instruments and Methods in Physics Research Section B: Beam Interactions with Materials and Atoms **511**, 1-5 (2022).
6. Inorganic Crystal Structure Database (ICSD) collection code: 632921.
7. L.R.B. Elton, D.F. Jackson, X-Ray Diffraction and the Bragg Law, Am. J. Phys. **34**, 1036 (1966).
8. L.R. Doolittle, Algorithms for the rapid simulation of Rutherford backscattering spectra, Nucl. Instrum. Methods Phys. Res. B **9**, 344 (1985).
9. M. Mayer, SIMNRA, a simulation program for the analysis of NRA, RBS and ERDA, in: J.L. Duggan, I.L. Morgan (Eds.), Proceedings of the 15th International Conference on the Application of Accelerators in Research and Industry, Vol. **475**, American Institute of Physics Conference Proceedings, p. 541 (1999).
10. MA Lampert, P. Mark, Current injection in solids, New York (NY): Academic Press, (1970).

Spanning the Periodic Table: select examples of stable isotope target fabrication at Oak Ridge National Laboratory

Cherlyn Foster[1*], *Jenny* Conner[1], and *Mike* Zach[1]

[1]Stable Isotope Materials and Chemistry Group, Oak Ridge National Laboratory, Oak Ridge, Tennessee, USA

Abstract. This paper reviews select examples of recent targets produced by the group, including ^{171}Yb wires, ^{62}Ni pellets, and ^{25}Mg foils. The targets discussed will cover the processes and equipment used for stable isotope manufacturing at ORNL and give a broader understanding of thin film preparation techniques, preparation of high-purity and specialty materials, and development of highly material-conservative methods.

1 Introduction

The US Department of Energy (DOE) Isotope Program's National Stable Isotope Repository at Oak Ridge National Laboratory (ORNL) is a DOE Office of Science capability that ensures a supply of crucial isotopically and chemically purified materials used for cancer therapy, diagnostic imaging, security, and scientific research.

The Stable Isotope Materials and Chemistry (SIMC) Group at ORNL manages the National Stable Isotope Repository and performs custom technical services to transform any material in inventory into the target form needed by the researcher. The group has a strong program history and continues to meet the needs of researchers [1, 2]. The capabilities of this group will be reviewed.

Specific fabrication techniques and equipment employed in recent target manufacture will be covered in detail including the equipment and methods required. The fabrication examples are ^{171}Yb wire from ^{171}Yb oxide powder, ^{62}Ni pellets from ^{62}Ni metal powder, and ^{25}Mg foil from ^{25}Mg metal piece. Specific yield information is not included for each method because these yields can differ between batches of isotopic materials. Procedures are optimized based on feasibility trials with natural materials and then applied to each isotopic batch.

2 Capabilities for Stable Isotope Dispensing and Fabrication at ORNL

The SIMC Group at ORNL specializes in stable isotope target fabrication techniques. The group is distinctive for a few reasons: the ability to perform transformations at a very small scale; the combination of materials, tools, and skills for executing specialty targets; and the ability to perform the work with near zero loss of the valuable isotopic material.

Isotopes can be ordered or inquired about by contacting the National Isotope Development Center (NIDC), and more information regarding the isotope catalogue and availability is found at isotopes.gov. Dispensing of stable isotopes and related business operations are conducted under an ISO 9001:2015 certified quality system.

The extremely high value of enriched stable isotopes necessitates that the manufacture of targets is performed with a minimal material loss during chemical conversions, fabrication of targets, and handling of enriched gases. Target forms can include powder, foil, disk, wire, pellet, bead, rod, or gas, depending on the isotope of interest.

The transformations to custom target forms include the following: (1) pyrochemical and thermal conversions; (2) inorganic compound synthesis; (3) powder metallurgy by pressing, melting, and/or sintering; (4) hot and cold rolling; (5) wire casting, rolling, swaging; (6) arc melting, casting, alloying, and drop casting; (7) ion beam and plasma sputtering; and (8) high-vacuum evaporation. These techniques are performed with a unique set of equipment combined

* Corresponding author: fostercj@ornl.gov

Notice: This manuscript has been authored by UT-Battelle, LLC, under contract DE-AC05-00OR22725 with the US Department of Energy (DOE). The US government retains and the publisher, by accepting the article for publication, acknowledges that the US government retains a nonexclusive, paid-up, irrevocable, worldwide license to publish or reproduce the published form of this manuscript, or allow others to do so, for US government purposes. DOE will provide public access to these results of federally sponsored research in accordance with the DOE Public Access Plan (http://energy.gov/downloads/doe-public-access-plan).

with analytical tools to ensure the target meets or exceeds customer expectations. The SIMC Group's physical location at ORNL provides opportunities for equipment and expertise beyond the group itself, which enables additional techniques and analysis.

3 Review of Fabrication Methods

3.1 Fabrication of ^{171}Yb Wire from ^{171}Yb Oxide

The ^{171}Yb wire is fabricated by converting the inventory form of ^{171}Yb oxide into a metal and then swaging to a wire. The starting material and final wire are shown in Fig. 1.

The process starts by drying the ^{171}Yb oxide powder in a muffle furnace to remove any moisture. The oxide is converted to metal by lanthanum reduction and distillation in a vacuum induction furnace (Fig. 2) [2]. The use of aluminium powder was reported as an alternate reductant for ytterbium oxide; however, lanthanum was used by others more recently [3, 4]. The distilled ytterbium was not tested for lanthanum impurities, testing for impurities and isotopic assay is done in the oxide form prior to fabrication.

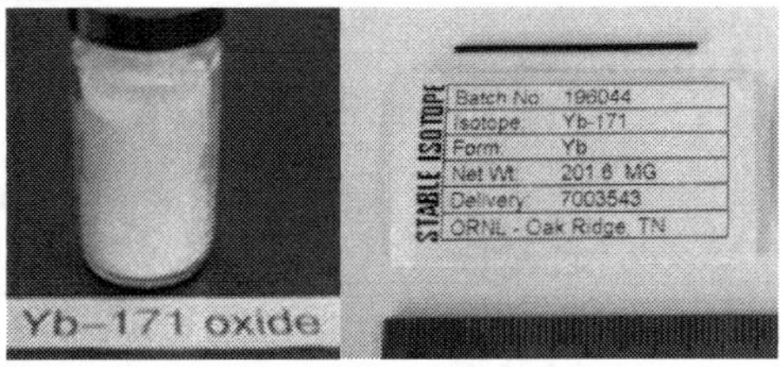

Fig. 1. The ^{171}Yb oxide inventory material and final ^{171}Yb metal wire.

Fig. 2. Operating the Denton Vacuum Furnace to reduce the ^{171}Yb oxide to metal and distill.

After the metallic ^{171}Yb is collected, it is melted using a Centorr arc melter to form a bead suitable for swaging. The arc melter and resulting bead are shown in Fig. 3 and Fig. 4.

The bead is manually swaged and then mechanically swaged to the final diameter desired by the customer. Fig. 5 shows the small mechanical wire swager used for wire fabrications.

Fig. 3. Operating the Centorr arc melter to melt the ^{171}Yb metal into a bead for swaging.

Fig. 4. The ^{171}Yb metal bead formed from arc melting.

Fig. 5. The Fenn wire swager.

3.2 Fabrication of ^{62}Ni Pellets from ^{62}Ni Metal Powder

The ^{62}Ni pellets are fabricated by pressing and thermally treating ^{62}Ni metal powder. The ^{62}Ni metal powder and a completed pellet are shown in Fig. 6.

The ^{62}Ni metal powder is initially reduced under hydrogen and argon gas mixture in a tube furnace at 700°C for 3 hours to remove any oxide that formed during inventory storage [5]. The tube furnace setup is shown in Fig. 7. This ensures that the material is a metal powder with an oxide layer that is as thin as possible, thereby causing better adhesion between particles.

Fig. 6. The ^{62}Ni metal powder inventory material and final ^{62}Ni pellet.

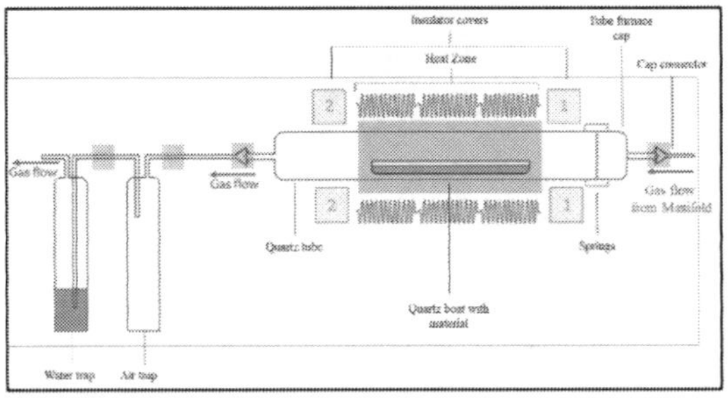

Fig. 7. A schematic of the tube furnace apparatus used for reduction and stearic acid removal.

After the ^{62}Ni metal powder cools from the reduction, it is carefully weighed out in amounts specified by the customer's desired pellet size. The powder is loaded into a die coated with stearic acid (i.e., to prevent adhesion to the die) and pressed in a manual hydraulic press (Fig. 8).

Fig. 8. Aligning the die to press a ^{62}Ni pellet and the Carver manual hydraulic press.

After the pellets are pressed, they are heated again in a tube furnace to remove the stearic acid from the pellets. The heating profile used is 30 minutes at 400°C under argon and then 1.5 hours at 700°C under hydrogen and argon mix. After they cool, the pellets are measured and delivered to the customer in a tray to maintain traceability to size and processing parameters. Fig. 9 shows a batch of completed ^{62}Ni pellets.

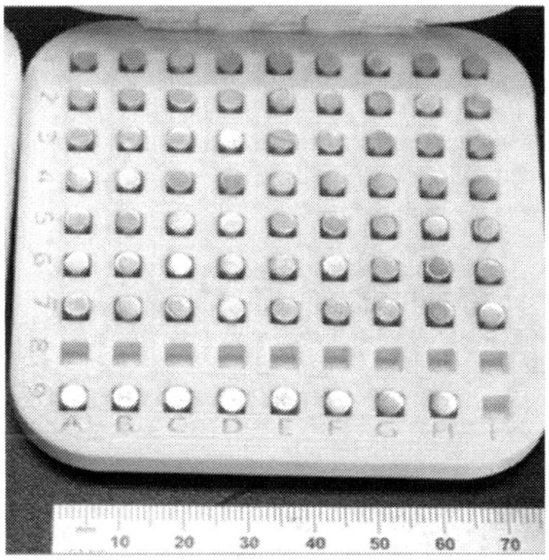

Fig. 9. Completed ^{62}Ni pellets ready for delivery to the customer.

3.3 Fabrication of ^{25}Mg Foil from ^{25}Mg Metal Piece

The ^{25}Mg foils are fabricated by rolling material cut from a piece of ^{25}Mg metal (Fig. 10). The fabrication normally starts with an available piece of isotopic magnesium in inventory; however, magnesium foils can also be obtained by starting with evaporated foils or melted or distilled material [2, 6, 7].

Fig. 10. The starting piece of ^{25}Mg from inventory and completed ^{25}Mg foils.

A small piece, which is sized based on the final size requirement of the foils, was cut from the large inventory piece of ^{25}Mg metal using a diamond wire saw to minimize losses. Material lost as kerf in this process caused by the saw blade width is recollected as an oxide powder for reprocessing and repurification by the SIMC chemists. This piece is welded into an argon-purged stainless steel pack for hot rolling (Fig. 11). Welding into a stainless steel pack prevents oxidation caused by repeatedly heating the material to 80–400°C for hot rolling, which is empirically determined by the material used and its thickness.

Hot rolling is a process that combines annealing and rolling techniques. It is used in cases wherein the material is too brittle to be rolled effectively by cold rolling, too thick to be rolled to the desired thickness in a timely manner, would require annealing throughout the process to roll, or the material itself requires a specific annealing process to be effectively formed into self-supporting targets. Hot rolling is rarely used in the

https://doi.org/10.1051/epjconf/202328506002

fabrication of metal targets if the starting metal piece is less than 1 mm thick because of the introduction of oxide and surface roughening that is inherent in the hot rolling process. However, for some select metals (e.g., magnesium), the process of hot rolling is required because of how quickly the metal becomes work hardened. If magnesium is not properly annealed at varying temperatures before cold rolling, then internal stresses will build to the point of shattering at less than a 30% decrease in thickness. Hot rolling allows this metal piece to gain the crystal structure necessary for effectively cold rolling the metal.

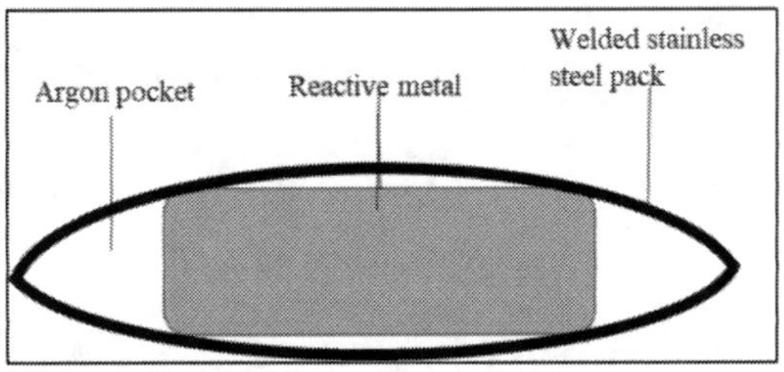

Fig. 11. A schematic of a typical hot rolling pack.

The hot rolling pack is heated and then rolled repeatedly until the desired size and thickness is achieved. The piece is monitored inside the pack using an X-ray image (Fig. 12).

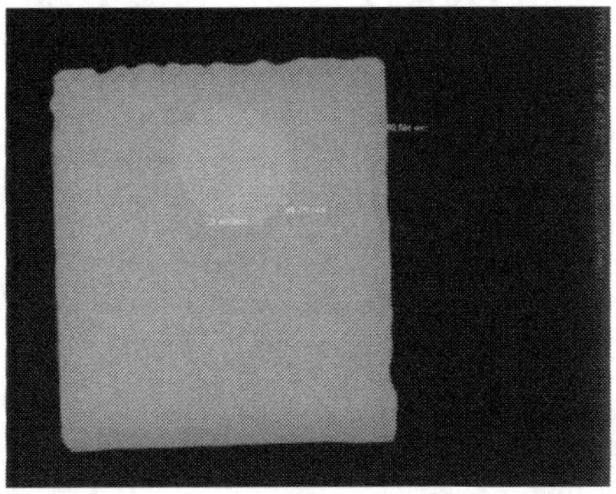

Fig. 12. An X-ray image of a metal piece inside a hot rolling pack indicating the dimension of the piece after several hot rolling iterations.

After the piece has been rolled to the desired thickness, it is removed from the pack and cold rolled to final foil thickness. It is placed inside a work-hardened stainless steel sleeve to allow for addition of a lubricant, such as ethanol or isopropanol, as well as for protection from direct contact with the rollers (Fig. 13).

The thin foil is susceptible to cracking and formation of pinholes and must be handled with extreme care. The foil is weighed and measured to determine when the desired thickness has been achieved. It is inspected using a light table to ensure the overall quality is as specified with respect to pinholes and then cut to the required dimensions for the customer.

Fig. 13. The ^{25}Mg inside the cold rolling sleeve, being rolled in the Fenn rolling mill.

4 Summary

The SIMC group at ORNL performs stable isotope material fabrications as highlighted in this report, with the unique combination of extremely low loss techniques and specialty equipment dedicated to this work. Many isotopes and product forms are available, and the group continually develops new techniques and methods as required by the customer.

This work is supported by the US Department of Energy Isotope Program, managed by the Office of Science for Isotope R&D and Production.

References

1. W.S. Aaron, J.G. Tracy, E.D. Collins, Nucl. Instr. and Meth. **A397**, 9 (1997)
2. L.O. Love, Proc. of the 1974 Conf, 29 (1974)
3. E.H. Kobisk, H.L. Adair, Nucl. Instr. and Meth. **167**, 153 (1979)
4. Clifford, S., Guo-ji, X., Ingelbrecht, C., Pomeroy, M.J., Nucl. Instru. and Meth., **480**, 29 (2002)
5. E.H. Kobisk, Report AERE-R **5097**, 103 (1965)
6. M. Heagney, J.S. Heagney, Report LA-6850-C, 92 (1977)
7. J. Jaklovsky, *Preparation of nuclear targets for particle accelerators*, 125 (Plenum Press, 1981)

Preparation and characterisation of High Oriented Pyrolytic Graphite backed targets for the NUMEN project

Antonio Massara[1*], *Maria* Fisichella[1], *Martina* Ursino[1], *Vittoria* Capirossi[2,3], *Francesco* Cappuzzello[1,4], *Manuela* Cavallaro[1], *Franck* Delaunay[1,4,5], *Federico* Pinna[2,3], *Alberto* Rovelli[1] *for the NUMEN collaboration*

[1]Istituto Nazionale di Fisica Nucleare, Laboratori Nazionali del Sud, Catania, Italy
[2]Department of Applied Science and Technology, Politecnico di Torino, Torino, Italy
[3]Istituto Nazionale di Fisica Nucleare, Sezione di Torino, Torino, Italy
[4]Dipartimento di Fisica e Astronomia "Ettore Majorana", Università di Catania, Catania, Italy
[5]LPC Caen, Normandie Université, ENSICAEN, UNICAEN, CNRS/IN2P3, Caen, France

Abstract. The equipment for target production and characterisation at the Laboratori Nazionali del Sud – INFN (Catania, Italy) is described. In particular, the CACTUS system used for target thickness and non-uniformity measurements is detailed. First trials in view of the production of Cadmium targets, under development for the NUclear Matrix Elements for Neutrinoless double β decay (NUMEN) project, are reported and discussed.

1 Introduction

The target laboratory at Laboratori Nazionali del Sud (LNS) of INFN at Catania has over thirty years of experience in target preparation for nuclear physics experiments and interdisciplinary physics. Users are not only local scientists but also researchers from other universities and laboratories worldwide. A list of the targets produced at LNS and their respective production techniques are given in [1].
In the present paper, the production and characterisation of a Cd target prototype, recently realised for the NUMEN project experiments [2-4], will be discussed in detail together with a brief description of the devices available at LNS for target preparation and characterisation.

2 LNS equipment for target preparation

Physical Vacuum Deposition (PVD) is a common method in the preparation of targets for nuclear experiments. PVD systems deposit thin films by a process in which a solid target material is vaporized, transported in vacuum and condensed on a substrate. PVD processes include thermal evaporation, electron beam (e-beam) evaporation, and sputtering. At the LNS laboratory, two evaporation systems are present: a classical L300 Thermal Bell Jar (TBJ) evaporator and a L560 Leybold-Heraeus, equipped, respectively, with a resistive source and an e-beam source. Thermal evaporation is typically used for materials which have a melting point lower than 1500 °C, whereas the e-beam gun is well suited for materials with high melting point temperatures and/or which could alloy or chemically react with crucible materials causing a contamination of the deposited film.

2.1 Thermal Bell Jar Evaporator

The L300 TBJ evaporator (Figure 1), with a power of 1.1 kW, has a stainless steel base and a glass bell jar with a diameter of 300 mm and a height of 350 mm.

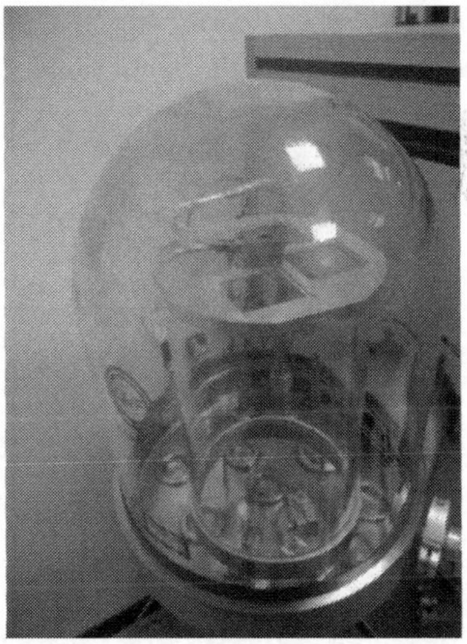

Fig. 1. View of the L300 Thermal Bell Jar evaporator. It is possible to distinguish the glass cylinder on which the target frame holder is placed and the quartz balance at the central upper position.

* Corresponding author: massara@lns.infn.it

 https://doi.org/10.1051/epjconf/202328506003

It is equipped with two pockets for material deposition. Source materials are melted by resistive heating in metal evaporation boats (Ta, W, Mo) or ceramic crucibles. Evaporation rates are manually controlled by varying the output power of a power supply. Deposition rates are monitored with a quartz crystal microbalance. Using glass cylinders of different heights, as shown in figure 1, it is possible to vary the distance between the source and the substrate (on which the evaporated material will be deposited) in a range from 80 to 220 mm.

2.2 L560 Leybold-Heraeus Low Vacuum Evaporator

The L560 evaporator at LNS (see figure 2) is equipped with a resistive heating source as well as with a 6 kW 270° electron-beam gun, with a 4 pocket evaporation source. It is possible to choose which one must be used according to the properties of the material to be evaporated. The evaporator is equipped with a quartz microbalance crystal and, as in the case of the L300 evaporator, it is possible to set the distance between the source and the substrate using glass cylinders of different heights. Moreover, the evaporator has a halogen heating element to set the temperature of the substrate, if needed. Indeed, as is well known [5], that heating the substrate during the deposition can support the diffusion of the material on the substrate surface, thus avoiding the formation of islands or typical structures. The quartz and the heating system are mounted on movable arms so that they can be coupled with both systems. In the bottom part of figure 2, one can see a typical set up for an e-beam evaporation (inside the red circle): the 8-cm^3 hearth with its cooling system (to maintains the source temperature lower than 100 °C), the microbalance and the heating elements. On the left, the electrodes for thermal evaporation are also visible.

Fig. 2. Internal view of the L560 evaporator. The setup is prepared for e-beam evaporation. Inside the red circle one can see the e-beam hearth, the quartz balance and the heating elements. On the bottom left part the electrodes for thermal evaporation are also visible.

2.3 Rolling mill

A cold rolling mill is also available for preparing metal targets with thicknesses greater than 1 μm. It consists of two hardened and polished stainless steel rollers of 125 mm face width by 85 mm diameter, driven by a motor. The metal foil is progressively thinned by passing between these two rollers until it reaches the desired thickness.

3 New system for target characterisation at LNS

The target thickness is checked during the evaporation using the quartz microbalance of the evaporator. To have a more accurate value of the single target thickness, Alpha Particle Transmission (APT) measurements are usually performed. Recently, the LNS target laboratory has been equipped with an apparatus named Chamber for Alpha-particle Characterisation of target Thickness and Uniformity by Scanning (CACTUS). The system, which uses the APT technique, was entirely developed at LNS. It is a cylindrical chamber of about 1 m in diameter containing four sectors where it is possible to place the targets to be characterised (see Fig.3). Each sector has been designed for different target frame geometries. The measuring system (see Fig. 4) consists of a moving probe with a special head containing a ^{241}Am α-source (55.5 kBq unsealed source) and an ion implanted silicon detector, 100 μm thick and with a diameter of 25 mm^2. In front of the source and the detector two collimators, respectively, of 6 mm and 3 mm of diameter are mounted. The distance between the two collimators is 33 mm, while the distance between a target and the detector collimator is about 6 mm. According to this geometry, the dimension of the alpha spot in the target will be about 3 mm. The head has been designed to allow the sector movement in the space between the source and detector.

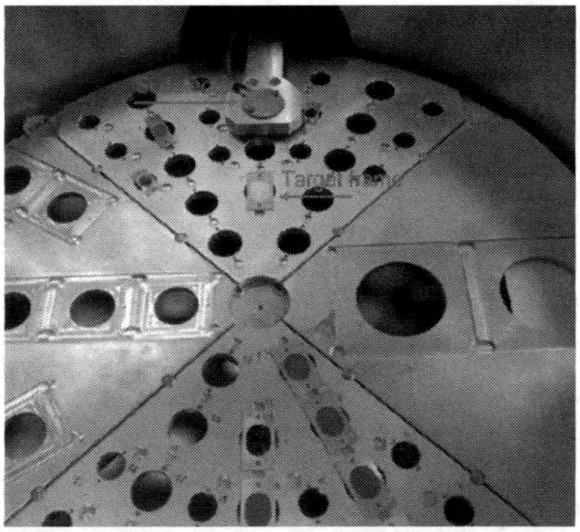

Fig. 3 Internal view of the CACTUS chamber. The four plates to place the target frames and the top part of the radial probe are visible.

The radial probe movement and the sector rotation allow positioning the head over the target to be characterised. A video camera, also located on the head, guarantees precise automatic target positioning with a precision within 1 mm. Control loop and automated procedures ensure the measurement of many targets without opening the vacuum chamber. The CACTUS system allows to investigate the target thickness on the whole target surface, making possible a study of the uniformity between different areas of the target (''global'' uniformity). An ORTEC/AMETEK easy-MCA multichannel analyser with the MAESTRO software is used for data acquisition.

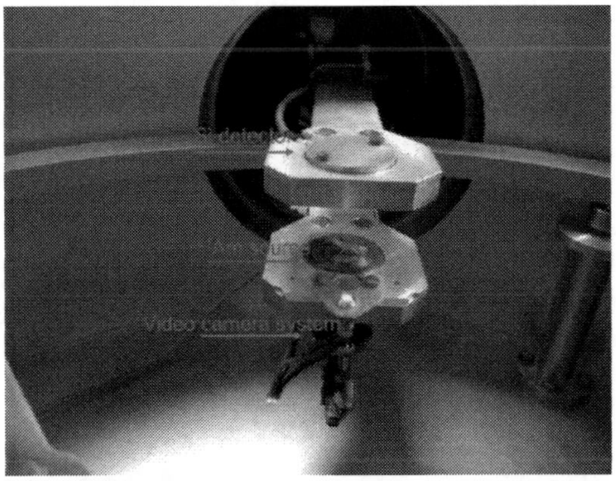

Fig.4 Close-up view of the measuring probe. The source and the detector are placed in the bottom and top part, respectively, of the moving arm. In the foreground the LED-video camera system is also visible.

The measured target thickness is the average value over a spot surface, defined by collimators placed in front of the source and the detector. In measurements reported in the following, the geometry of the system was such that the alpha spot on the target was about 4 mm.

4 Targets for NUMEN experiments

In this section, preliminary results concerning the Cadmium target production for the NUMEN (Nuclear Matrix Element for Neutrinoless double beta decay) experiments [2-3] are reported. Targets for these experiments will be made with isotopes suitable as candidates for neutrinoless double beta decay studies (for example ^{48}Ca, ^{76}Ge, ^{76}Se, ^{82}Se, ^{96}Zr, ^{100}Mo, ^{106}Cd, ^{116}Cd, ^{110}Sn, ^{124}Sn, ^{128}Te, ^{130}Te, ...). They have to be deposited on High Oriented Pyrolytic Graphite (HOPG) [6], which thanks to its high thermal conductivity (1950 $Wm^{-1}K^{-1}$) could help dissipating the high power released by very intense ^{18}O and ^{20}Ne beams up to 10^{13}pps used for these experiments. According to thermal calculations, the minimum value of HOPG for efficient heat dissipation is about 450 μg/cm^2. To maintain a reasonable resolution on the energy of reaction products, the target thickness will range from 250 to 500 μg/cm^2, depending on the material to be evaporated. Moreover, both the backing and the target have to be uniform, to avoid degrading the energy resolution on the reaction products by additional straggling effects. For this reason, the APT measurements are also performed on the backing before the evaporation process to determine its thickness and uniformity. Furthermore, the characterisation of the HOPG substrates provided by different companies are being characterised and the results will be presented elsewhere. In the present report, the focus will be on the cadmium target manufacture and characterisation.

4.1 Cadmium target production

The required thickness for the cadmium layer deposited films was about 500 μg/cm^2. Natural cadmium material was used, not to waste isotopically enriched material. The evaporation was performed at LNS using the L300 evaporator with the following working conditions and geometry:

- 1.5 - 2g Cd deposited on Mo boats;
- distance source - backing equal to 250 mm;
- Evaporation rate: 0.2 Å/s.

At the end of the evaporation procedure, about 1g of Cd was used and 8 targets with an area of 12 mm x 15 mm were produced.

In the same evaporation both 450-μg/cm^2 thick HOPG, and thin (70 μg/cm^2) amorphous carbon, manufactured by OPTIGRAPH and Yissum respectively, were used as backings to study the possible effects due to different substrate. Since it is well known that the cadmium does not stick well on carbon substrates, a 0.1 μg/cm^2 bismuth buffer was deposited on the backing to improve the cadmium adhesion. The first trial did not give good results: The cadmium thickness measured by APT resulted thinner than the expected one, measured by the quartz microbalance. The average thicknesses of the targets produced are reported in the first two rows of Table 1. A different behaviour has been observed if the backing was an amorphous carbon foil. The amount of cadmium material deposited on pyrolytic graphite is about half of the one deposited on amorphous carbon. Literature [3] suggests that the best result for evaporated Cd targets is reached if Cd is deposited on a cooled backing.

Evaporation procedure	Backing-Type	Cd thickness (μg/cm^2)
Single-step	HOPG (450 μg/cm^2)	170
Single-step	C (70 μg/cm^2)	300
2-step	HOPG (450 μg/cm^2)	140
2-step	C (70 μg/cm^2)	474
2-step	HOPG (2.25 mg/cm^2)	370

Table 1. Evaporation procedure (see text for details), type and nominal thickness of the backing (from the manufacturer), and average thickness of the cadmium target obtained.

Since the evaporator systems at LNS do not offer the possibility to cool the substrate, a new trial was performed by evaporating the Cadmium in two steps. In the first one, the cadmium material is evaporated till the half of required amount is deposited. Then follows an intermediate phase to reach room temperature before the second cadmium evaporation step to reach the final thickness. The deposition thickness is monitored using the evaporator quartz crystal microbalance. The starting conditions were identical to the ones described above. In this second target series, in addition to HOPG and amorphous C backings similar to those used in the first test, 2.25-mg/cm^2 thick HOPG, substrates, produced by PANASONIC, were used.

In the last three rows of Table 1, results obtained in this 2-step evaporation are reported. The thickness of the cadmium evaporated on the 450 μg/cm^2 HOPG is similar to the ones of the single-step procedure. On the contrary, the two-step procedure improved the deposition on amorphous carbon: A result very close to the expected one has been obtained. Concerning the third type of substrate, the 2.25-mg/cm^2 thick HOPG, a thicker deposition has been obtained compared to the case of a thinner HOPG substrate, but the thickness is still lower than the expected one. This seems to indicate that the surface of the foil plays an important role in the deposition process. Further studies are necessary to understand the different behaviour of cadmium on C and HOPG. In particular, scanning electron microscopy analysis to study the backing surface has already been planned.

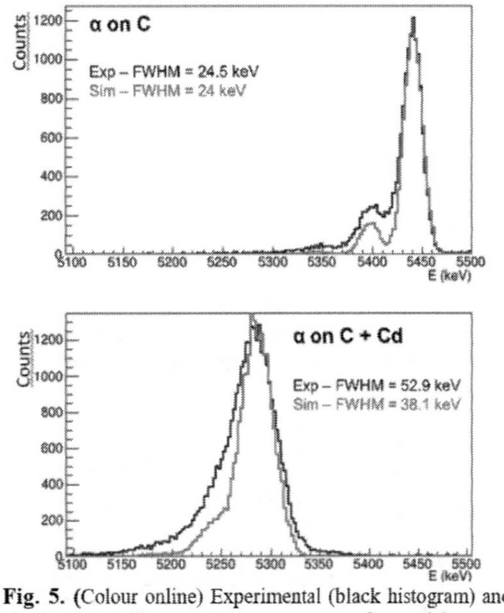

Fig. 5. (Colour online) Experimental (black histogram) and simulated (red histogram) energy spectra of α-particles passing through an amorphous carbon backing alone (top) and through a C+Cd target (bottom). The simulated spectra are obtained with the SRIM [8] code.

For the targets deposited on the amorphous carbon, with the expected thicknesses, a detailed analysis has been performed to study the uniformity of the cadmium deposition. Thanks to the CACTUS system, the thickness has been measured in several regions across each target, to investigate its dependence on the position: Differences inside 2% have been observed between the measured points, confirming a good global uniformity of the deposition. The local uniformity (i.e. the uniformity inside the area covered by the alpha-particle spot, 3mm) of the cadmium layer was extracted following the procedure described in [7], by comparing the Full Width at Half Maximum (FWHM) of the experimental spectrum with the FWHM of a spectrum simulated assuming a uniform target with the measured average thickness. In figure 3, the residual energy spectra for alpha-particles crossing the carbon backing alone (top) and the C+Cd target (bottom) are reported. The black histogram is the experimental spectrum, and the red histogram is the result of a SRIM simulation [8]. The local non-uniformity of the backing has been estimated to be less than 10%, while for the cadmium layer a local non uniformity of about 25% has been determined.

5 Conclusions

The devices available at the LNS target laboratory allow to produce several types of targets. The new CACTUS system allows a complete characterisation of the target in term of thickness, local and global non-uniformities. Obviously, the development of new targets requires multiple trials and in addition to APT other types of analysis can help to better understand the deposition process. Examples of very useful analyses are given by scanning electron microscopy, to study the morphology of the sample surface, and Rutherford backscattering spectroscopy, to investigate the purity of the target.

Regarding the development of cadmium targets discussed in this report, further studies are necessary to obtain targets with the required thickness. A particular analysis must be devoted to the surface of the HOPG backing.

References

1. https://www.lns.infn.it/en/targets.html.
2. F. Cappuzzello et al, Eur. Phys. J. A (2018) **54**: 72
3. F. Cappuzzello et al. International Journal of Modern Physics A 36, (2021) 2130018
4. P. Finocchiaro et al. Universe 6 (2020) 129
5. Stolarz, J. Radioanal Nucl. Chem. (2014) 299:913-931
6. F. Pinna, Acta Polonica B, 51, 3 2020
7. M. Fisichella, A. Shotter et al., Phys. Rev. C **92** 064611 (2015)
8. J. Ziegler, "SRIM code" [http://www.srim.org]

Design of a high power solid target for ^{211}At

Guillaume *Brunet*[1*], Michel *Matthieu*[1], Gilles *De France*[1], Vincent *Desmezières*[1], Bertrand *Jacquot*[1], Clément *Hocini*[1], Sébastien *Ferey*[1], Jérôme *Gueret*[1], Laurent *Rossard*[1], Martial *Morisset*[1], Georges *Fremont*[1], Sabrina *Lecerf*[1], Thierry *André*[1], Clément *Michel*[1], Patrice *Gangnant*[1], Jean-Claude *Foy*[1], Romuald *Levallois*[1], Cécile *Barthe-Dejean*[1], *and* Matthieu *Collard*[1]

[1]GANIL Bd Henri Becquerel, 14076 CAEN Cedex 05 FRANCE

Abstract. The new SPIRAL2 LINAC at GANIL delivers light beams (p, d, alpha,...) with very large intensities. In order to exploit these beams in an optimal way, high power target stations are requested. In this contribution, we present such an irradiation station designed to optimize the production of ^{211}At, an alpha emitter of interest for targeted alpha therapy. In order to dissipate at least 10 kW of beam power, a rotating, water-cooled system comprising twelve individual targets has been developed together with a fast and safe extraction mechanism.

1 Introduction

While the use of radionuclides for diagnostic purposes (Single Photon Emission Computed Tomography, SPECT, and Positron Emission Tomography, PET) is a mature field in nuclear medicine, the therapeutic use of alpha emitting radionuclides is still less evolved. Targeted Alpha Therapy aims at using locally highly cytotoxic properties of alpha particles by carrying the alpha decaying radionuclides to specific sites of cancerous cells. This would allow maximize the efficiency of the treatment while reducing damages done to healthy tissues.

^{211}At is considered as a promising candidate for such application thanks to its adequate half-life of 7.2 h and its alpha decay mode [1]. Currently, an efficient way to produce this radioisotope is the reaction ^{209}Bi (^{4}He, 2n) ^{211}At. As the needs for ^{211}At for research or future medical applications are bound to strongly increase [2], up-to-date accelerators, including linear accelerators, able to deliver alpha beam at high intensities with precise energy and high quality bismuth (Bi) targets are mandatory.

To address this challenge, the REPARE (Research and developments for the Production of innovAtive RadioElements) project aims at improving ^{211}At availability by, amongst other ways, developing technologies to improve production capacities through dedicated facilities [3, 4]. One specific task of this project is the development of specific targetry able to sustain high intensity beams either considering liquid (cf T. Bigourdan et al in this proceeding) or solid target, developed in this paper.

2 ^{211}At production

The radioisotope of interest, ^{211}At, is produced by fusion-evaporation reaction using bismuth targets irradiated with an alpha beam at an energy higher than 20.7 MeV. The excitation function to produce ^{211}At has a maximum at about 30 MeV. However, this energy is above the threshold energy of reactions leading to cross-contaminants such as ^{210}Po, decay products of the three-neutron channel (^{210}At), which is problematic. Indeed, ^{210}At decays with a half-life of 8.1 h to ^{210}Po, which is highly cytotoxic because of its huge specific activity ($1.66\ 10^{14}$ Bq/g) and long half-life (138.4 days). An efficient way to minimize the production of ^{210}At is to keep the alpha beam energy below the threshold of the ^{209}Bi (α, 3n) ^{210}Po reaction, i.e. 28.6 MeV. Therefore, the effective beam energy window to produce ^{211}At ranges from 28.6 down to 20.7 MeV [5]. With an ^{4}He beam in the mA range delivered by GANIL-LINAC, an end-of-beam activity of the order of 10 GBq/hour is reached and the targets should thus sustain a beam power of at least 10 kW. This is the main design specification of the REPARE irradiation station, which is described in the following sections.

3 Overview

3.1 Main design specifications

The challenge for the design of the solid target station is therefore to dissipate in a very efficient way the incident beam power in order to maintain the Bi in its solid state all along the irradiation time. The Bi material is chosen as metallic, with a melting point of 271°C, instead of oxide or sulphur form to prevent any by-products and to make easier the extraction of astatine by dry process.

* Corresponding author: brunet@ganil.fr

 https://doi.org/10.1051/epjconf/202328507001

These targets are made of a 70 µm evaporated Bi layer of 40*106 mm² area on a 3 mm thick substrate of aluminum nitride (AlN, area 50*120 mm²) for this study where the beam stops at 200 µm depth. The beam has a double-gaussian shape of 106 mm height and 2 mm wide. To reach the 10 kW dissipation goal, with 3 kW loss in Bi and 7 kW stopped in the substrate; the targets are mounted on a rotating water-cooled wheel hosting 12 targets. They are arranged in pairs on a structure, called a "racket", in order to minimize radiological exposure in compliance with an ALARA approach during their extraction from the wheel. The dimensions of the target, have been designed in such a way that they cover the large beam spot and can be easily handled for the subsequently chemical process to extract ^{211}At (Figure 1).

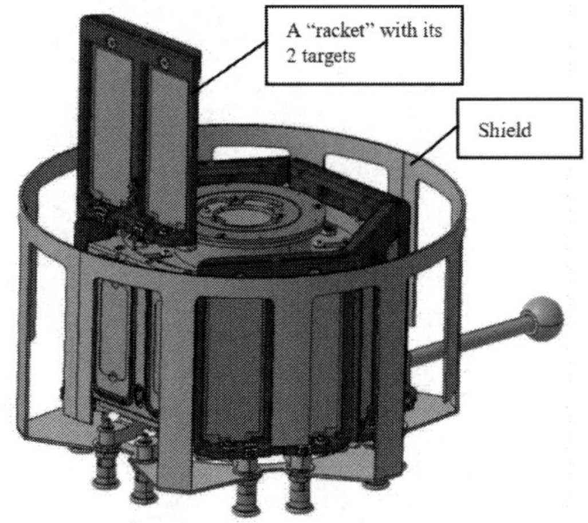

Fig. 1. Target wheel with one "racket" partially extracted

3.2 Target Station assembly

The REPARE irradiation station is composed of six main components listed in Figure 2. Its total weight and volume are 650 kg and 3x0.6x2.5 m3, respectively.

The assembly is positioned on a welded steel frame fixed on ground designed to withstand an earthquake, a compulsory requirement for each heavy component of the GANIL-SPIRAL2 facility.

The translation set composed of pneumatic actuators, supports the vacuum chamber, which can be placed either in alignment with the beam axis or out of it to enable irradiation of other targets placed upstream.

The motorisation set composed of a brushless motor, its encoder and dedicated mechanical reducer is connected to the perpendicular coaxial shaft of 50 mm diameter. To sustain the harsh radiation environments no electronic components are used.

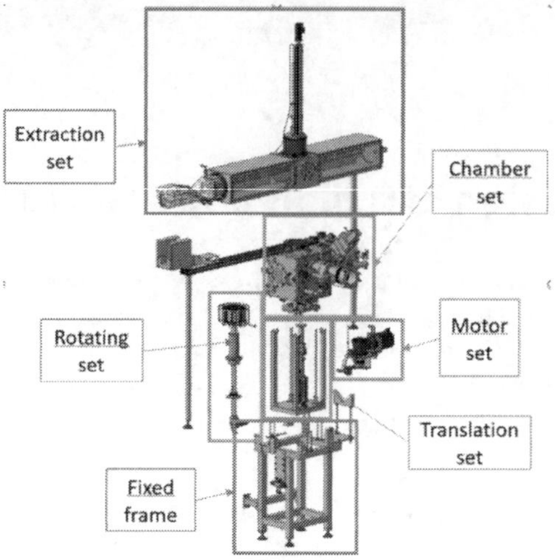

Fig. 2. REPARE irradiation station assembly showing the six main components

The rotating set includes a shield (s. 3.3.2), the wheel with its rackets, the above mentioned coaxial shaft, the rotating (ferrofluidic) vacuum-tight feedthrough and water cooling system (s. 3.3.1). This coaxial shaft drives the wheel and is made of a double tube for water inlet and outlet (Figure 3).

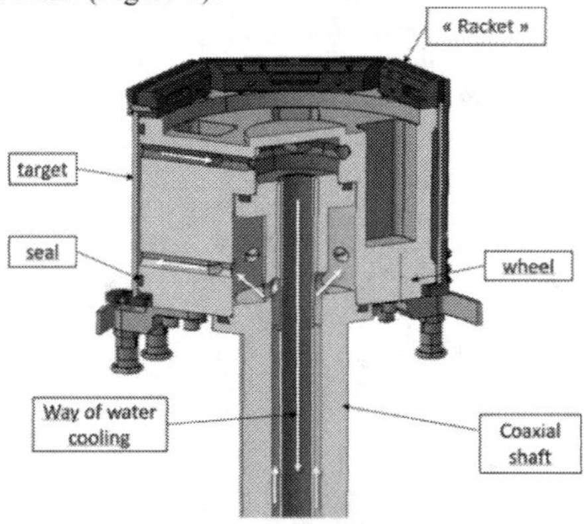

Fig. 3. Water cooling system: the cold water is cooling directly the rear surface of the target and is then extracted via the central part of the rotating axis.

The vacuum chamber is made of stainless steel with walls of 20 mm thick and opening of 30 mm thick to protect workers from radiations (below 45 µSv integrated dose) and is equipped with viewports and pumping systems to reach10^{-6}mbar.

The extraction set connected above the chamber is acting as a glove box. The targets extraction system must be reliable as well as ensuring a protection of the personnel in charge of the station operation against radiological exposure. At the end of beam time, the rackets (i.e. two targets) are extracted from the wheel using a clamp, which grasps the racket out of the wheel and store them into an individual lead container. Six such lead containers are installed in a long box ensuring double confinement of the generated activity (Figure 4). The lead containers are extracted from the station in a vinyl bag and stored in a dedicated parcel for shipment.

https://doi.org/10.1051/epjconf/202328507001

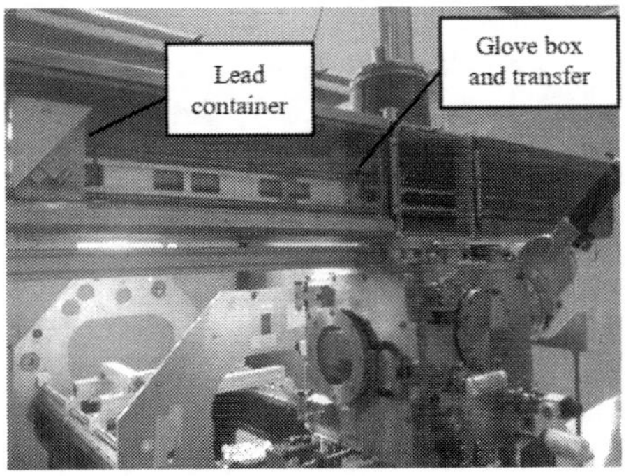

Fig. 4. Rackets extraction system

3.3 Modeling and functional parameters setting

To distribute the deposited beam power over large surfaces, the parameters of the cooling water (flux, and pressure) have to be quantified properly to prevent any risk of material melting and leakage. Moreover, precise measurements of beam intensity had to be developed in order to optimize the beam size and to synchronize it with the wheel rotation. These aspects are discussed below.

3.3.1 Fluidic simulations

For high power thick targets, the most efficient cooling method is thermal convection. Indeed the energy deposited by the beam inside the target and the substrate is transferred to the fluid flowing along the rear face of the later (AlN). However, the substrate has to sustain the water pressure while maintaining the sealing to prevent any leakage. In addition, the water flow has to be uniform over the rear face and all targets in order to guarantee a uniform surface cooling without any air bubbles. CFD (Computational Fluid Dynamics) simulations, taking into account rotation and cooling circuit conditions have been performed to evaluate the range of acceptable water pressure, its flux and rotation velocity. The cooling system (Figure 3) is modelled with ANSYS Workbench and FLUENT [6].

With a loading of 10 kW beam and a water pressure of 2 bars, the temperature of the target was calculated as a function of time. As is illustrated in Figure 5 for a rotation of the wheel at 100 rpm, the temperature is increasing in few seconds and stabilized to about 41°C. The temperature gradient is less than one degree at each beam impact. A simulation with a 400 rpm velocity indicates a final temperature of 31°C. This result is very conservative and makes possible to decrease the rotation velocity or the flow rate.

In addition, the velocity distribution of water along distributed inlets and rear face of the substrate was simulated at 200 rpm and 2 bars conditions. As shown in Figure 6, the water velocity is higher than 1m/s everywhere, sufficient to fulfill the design specification.

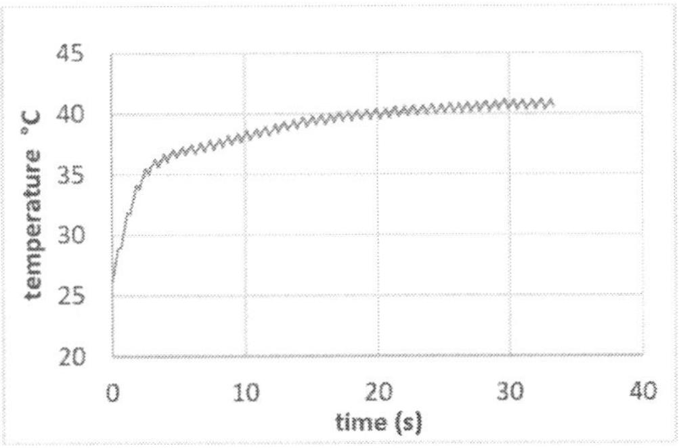

Fig. 5. Temperature evolution at 10 kW, 100 rpm, 2 bars

Fig. 6. Water velocity simulations at 10 kW, 200 rpm, 2 bars

3.3.2 Current measurements and beam setting

As mentioned earlier, it is crucial to minimize beam power deposition outside the effective target material (Bi), to measure precisely the beam current on target and to synchronize the time structure of the beam with the rotation of the wheel. Therefore, we designed a shield around the targets' wheel and developed an electronic system for current measurements.
Considering the aperture of the 12 targets of 20° each, a shielding structure was designed to protect the remaining 10° between two rackets from irradiation with stainless steel rolled sheets as depicted in Figure 7.

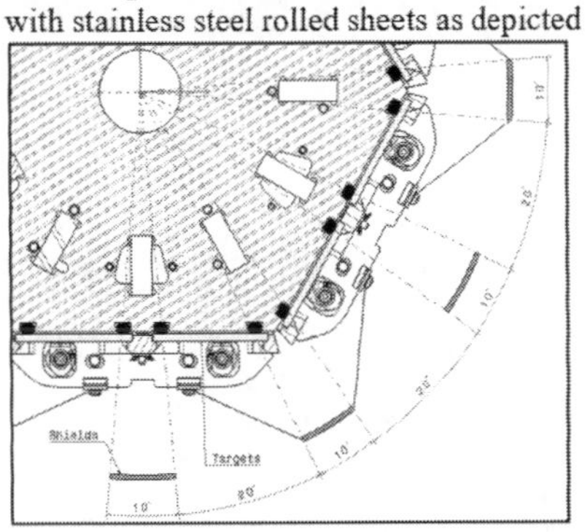

Fig. 7. Cross section of the rotating set across the targets (yellow) and the shield (orange)

This shield is electrically insulated from the wheel to enable current measurement of the eventually deposited beam, which is expected to range from 100 nA to 10 μA. With this current measurement and a setting of a periodical signal (20% on /10% off) in accordance with the velocity of the wheel, a delay on this signal can be adjusted and set to synchronize the beam with the targets.

To control the input intensity of the beam on the targets expected to range from 5 to 500 μA, current signals from the twelve targets are collected from a wire placed at the bottom of each to a connector placed in the middle of each "racket", named "touch pens" (Figure 8).

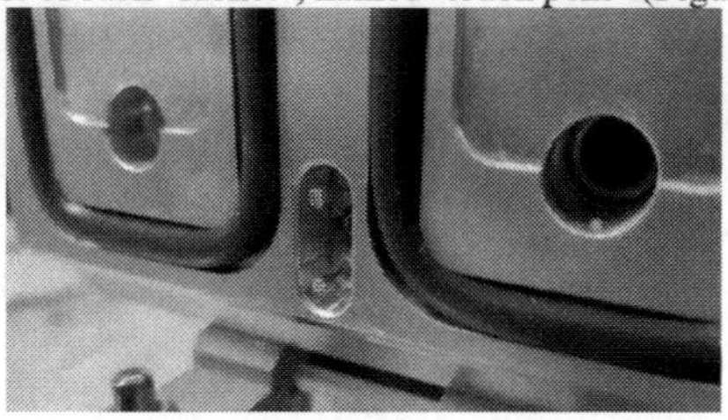

Fig. 8. "Touch pen" of each "Racket"

The signals are then routed inside the wheel to its top on a copper ring. The signals from the shield also routed to another concentric copper ring (Figure 9). These rings are in contact with two fixed brushes out of graphite to collect both signals.

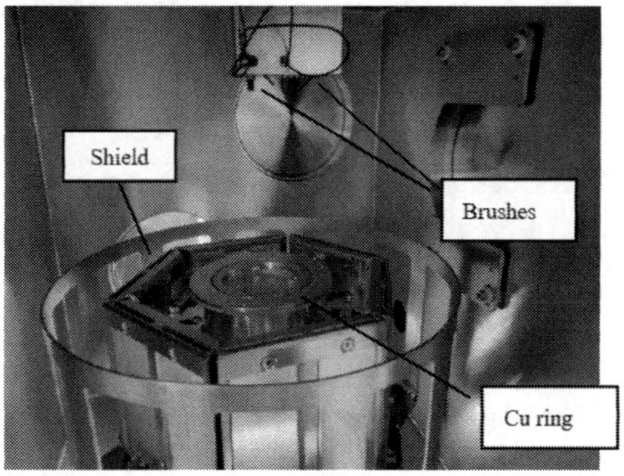

Fig. 9. Inside of the chamber (targets, measurement tools)

3.4 Off-beam tests

Following the detailed CFD simulations, the current measurement method and the reliability and reproducibility of the rackets extraction, several tests have been conducted. These tests consisted in:

• Controlling the absence of water leak at the level of the sealing of the targets with the wheel under 2 bars of water-cooling and 500 rpm of the wheel. The sealing is made of fluoroelastomers from Viton ® [7] and is radiation hard [8].

• Testing the water cooling efficiency and measuring its flow. Actually, the heating of the targets was simulated with a heating system of 150 W covering the substrate and thermocouples. With an inlet water at 22°C, a water flow of 50 l/min (i.e. about 4.4 l/min for each target) and a pressure of 2 bars, less than 50°C was measured on the substrate. Moreover, after switching off the water cooling and heating the target up to 180°C, the cooling system has been switched on again and the temperature decreased down to 70°C in 4 minutes.

• Measuring currents under rotation, up to 500 rpm, by injecting a low current of 10 nA, generated by a voltage and a resistance connected to the rotating copper ring. The collected current on the carbon brush was measured by an ammeter and visualized on an oscilloscope; it was stable and corresponds to the injected current with an accuracy of 1%.

• Validation of the extraction system and process handling, was revealed reliable, easy and robust.

4 Conclusion

A high power irradiation station has been designed in the context of the REPARE project, which aims at producing up to 100 GBq of ^{211}At in 8 hours of beam time using the very high beam intensity available at the GANIL SPIRAL2 facility. The target station is conceived to dissipate up to 10 kW of beam power. A cooling system combining direct water-cooling and target rotation has been developed. It has been tested offline and demonstrated to be very efficient. A dedicated current measurement method and beam synchronization with wheel rotation have been implemented. A reliable extraction system ensuring an easy and safe manipulation of the rackets has been designed. Detailed fluidic simulation calculations have been performed during the design process to ensure efficient cooling. Several offline tests focused on the absence of water leak during rotation, cooling efficiency, beam current reading and racket extraction mechanism have been conducted to confirm calculations and main functionalities. Currently, the REPARE irradiation station is assembled. The cooling efficiency, current measurements and beam synchronization will be tested under realistic conditions with beam in 2023. The first ^{211}At production run might be planned in fall 2023 provided the tests are all successful.

Acknowledgement

This work is supported by the French Research National Agency under the contract ANR-19-CE31-0013-01.

1. F.G. Kondev, M. Wang, W.J. Huang, S. Naimi, and G. Audi, Chin. Phys. C**45**, 030001 (2021)
2. F. Guérard, J.-F. Gestin and M. W. Brechbiel, Cancer Biother Radiopharm 28(1), 1-20 (2013)
3. F. Haddad, Eur. J. Nucl. Med. Mol Imaging **35**, no. 7, 1377-1387 (2008)
4. S. Galès, Nucl. Phys. **A834**, 717c-723c (2010)
5. I AEA, *Cyclotron Produced Radionuclides: Physical Characteristics and Production Methods, Series,* Technical Reports ed., vol. 468, Vienna: INTERNATIONAL ATOMIC ENERGY AGENCY (2009)
6. Ansys® Academic Research Mechanical, Help System, ANSYS, Inc., Release 2020 R2
7. https://www.viton.com/fr/applications/static-seals
8. H. Schönbacher, *Résistance aux rayonnements ionisants de matériaux installés autour des Accélérateurs du CERN*, CERN-HS-RP-072-CF (1982)

The PSI meson target facility and its upgrade IMPACT-HIMB

Daniela Kiselev[1*], *Pedro* Baumann[1], *Pierre-André* Duperrex[1], *Malek* Haj Tahar[1†], *Sven* Jollet[1], *Andreas* Knecht[1], *Daniel* Laube[1], *Davide* Reggiani[1], *Jochem* Snuverink[1], *Raffaello* Sobbia[1], and *Vadim* Talanov[1]

[1]Paul Scherrer Institute, 5232 Villigen, Switzerland

Abstract. The high intensity proton accelerator complex (HIPA) at the Paul Scherrer Institute (PSI) delivers a 590 MeV CW proton beam with currents up to 2.4 mA (1.4 MW). Besides two spallation targets for thermal/cold neutrons (SINQ) and for ultracold neutrons (UCN), the beam feeds two meson production targets Target M and Target E. The targets consist of graphite wheels of effective thickness 5 mm (M) and 40/60 mm (E). The target stations M and E are of quite different design; however, both of them rotate at 1 Hz to dissipate the heat (20 kW/mA for the 40 mm target E) efficiently. Recent progress was made by a new type of bearings and a new target geometry able to increase the rate of surface muons by up to 50 %. This is also foreseen for the upgrade of the target station M within the High Intensity Muon Beam (HIMB) initiative aiming to increase the surface muons available for experiment by two orders of magnitude. HIMB is part of IMPACT (Isotope and Muon Production with Advanced Cyclotron and Target Technology), an application for the Swiss Roadmap of Research Infrastructure.

1 Introduction

The 590 MeV proton beam of HIPA [1] with currents up to 2.4 mA serves four target stations, two spallation targets and two meson production targets. The Ultracold Neutron Source UCN gets the full beam for a few seconds every 5 minutes, whereas the neutron spallation source SINQ is located at the end of the main beam line after the two meson production targets M and E [2]. Therefore, the beam on the SINQ target already lost part of the energy and current on its way through target M and E. While this effect is negligible for target M due to the small effective target thickness of 5 mm graphite, the influence of target E (either 40 or 60 mm thick graphite), on the beam is substantial. E.g. after the 40 mm graphite the beam is reduced in energy by 15 MeV and by 30 % in current, mainly due to multiple scattering in the target. The positive side effect is that the emittance of the beam is already enlarged and therefore the beam can be distributed easily on a large area onto the SINQ target to avoid overheating. On the other hand, this also means that target E must not be missed any time to avoid damage of the SINQ target by a pencil beam.

Both targets are made of graphite, which has a large production rate for pions in the energy range of HIPA for a given beam transmission. Detailed studies [3] revealed that beryllium would be more effective. However, beryllium could not stand currents higher than 120 μA for more than a few weeks in the 590 MeV proton beam at PSI [4]. Most of the experimental stations at the seven beam lines (s. Fig. 1) prefer muons, which originate from pion decay. Muons are produced either by decay-in-flight of pions ('cloud muons') or by

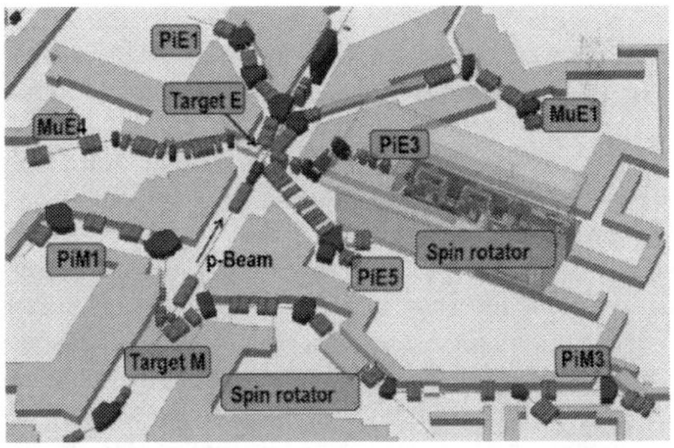

Fig. 1. Sketch of the five secondary beam lines around the Target E station and the two beam lines for muons and pions at the Target M station. Courtesy of Thomas Rauber, PSI.

stopped pion decay in the target. Muons produced close to the target surface, 'surface muons', leave the graphite, while those deeper inside are stopped. The large polarization of surface muons is used e.g. for investigating the magnetic structure of materials via the μSR technique [5]. With spin rotators, the spin of the surface muons can be turned in the desired direction. At PSI, three spin rotators are in use, the third one in PiE1 is not shown in Fig. 1. At the beam line MuE4 with a rate of the order of 10^8 surface muons per second, up to now the world highest intensity, low energy muons (LEM) of 0.5 to 30 keV are available.

2 Target E station

To operate a meson production target of 40 mm graphite in a 2.4 mA proton beam of 590 MeV and a size of about

* Corresponding author: Daniela.Kiselev@psi.ch
† now at Transmutex, Geneva, Switzerland

1.5 mm (2-sigma) poses several challenges. At such conditions, a power of 50 kW is lost at the target, which heats up the target and has to be dissipated. The material needs not only to be temperature resistant but also has to tolerate thermal stress as well as the change of physical properties due to radiation damage for a reasonable time period. In addition, not only the target material but also the mechanical parts of the target station have to withstand the radiation.

Natural graphite has a layered structure and therefore is quite anisotropic, i.e. its physical properties are significantly different perpendicular to and within the layer. E.g. when natural graphite is heated, large thermal stress occurs due to the different thermal expansion coefficients. Usually, this leads to cracks, if the thermal stress exceeds the ultimate tensile limit. To mitigate this, polycrystalline graphite is used instead. This graphite consists of very small grains, which have no preferred order and therefore have almost isotropic properties. Nowadays, the grains can be as small as 10 μm or even less. The smaller the grains the higher the density, which is achieved during sintering by isostatic pressing. Typically densities are between 1.8 and 1.85 g/cm^3, which is still 20% below the natural graphite. The grade used for Target E, R6510 from the SGL group, has a density of 1.84 g/cm^3 and a grain size of 10 μm.

The graphite target is a wheel with a diameter of about 450 mm, which rotates at 1 Hz to distribute the power on a rim with a circumference of about 150 cm. Due to the rotation, direct cooling by contact to e.g. water tubes would be difficult. In addition, since the thermal conductivity of graphite is known to decrease fast due to radiation damage [6], the cooling efficiency would drop soon. Therefore, the graphite is cooled by thermal radiation, which is an efficient way for materials with large emissivity at high temperatures. With an emissivity of 0.75 [4], the temperature of the graphite at the impact location of the beam is predicted to be 1700 K by simulations, since no direct measurement is available. The thermal power lost by radiation is dissipated by blocks of copper, which are cooled by water. One block is mounted directly behind the graphite wheel, others protect the vacuum chamber. Even though the vacuum chamber is cooled by water, it cannot dissipate as much power as the copper blocks, since the chamber is made of stainless steel with smaller thermal conductivity.

The wheel is connected to the hub through six spokes. At 1700 K the graphite wheel expands, which would lead to thermal deformation and cracks. For this reason the spokes can slightly slide on the wheel, such that the wheel is allowed to expand and cracks are avoided. However, with increasing beam current it was observed that the deformation of the wheel increases, which has to be limited to a maximum of 2 mm, since the target wheel is 6 mm wide only. This was significantly improved by choosing a graphite grade with more isotropic properties, i.e. smaller grains. In addition, the thermal stress was released by dividing the rim of the wheel in 12 segments with 1 mm space in between. The highest total proton charge on a target was 39 Ah, which would correspond to 4 to 5 full beam periods. However, the target was not continuously in use. After usage, the parts of the graphite containing a high concentration of desired radionuclides like ^{10}Be were used to prepare targets for new experiments [7,8].

The wheel is turned by a motor, which is mounted 2.5 m above the beam line to avoid radiation damage. At this location, it can be in operation for about 5 to 8 years. Unfortunately, the three ball bearings are mounted close to the graphite wheel and therefore in a high radiation field. That means that the bearings cannot be lubricated with grease, as an organic material would soon get hard and brittle, preventing a reliable operation. In addition, the first bearing can heat up to 140 °C, due to thermal radiation from the wheel but also due to the connection of the hub to the wheel by the spokes, which are hollow to minimize thermal conductivity. The demand that the bearings have to work under cold conditions as well as during beam operation requires a larger radial clearance in the bearings to compensate for thermal expansion. Both effects reduce the lifetime of the bearings.

When the bearings fail they have to be replaced. For this, the 4 m thick shielding of many concrete blocks has to be removed to access the so called service platform. The water supply is disconnected from the target insert and the remaining water removed. The vacuum in the beam line is broken, however, to minimize tritium release, an underpressure is kept. Due to the high dose rate of the target of about 3 Sv/h mainly due to ^{7}Be, the target insert has to be taken out by a 45 t exchange flask shielded by up to 40 cm of steel. The flask is remotely operated by a control unit showing the operation status of the flask on a screen. Due to the infrastructure on the service platform above the beam line, there is no space to position the flask. Therefore, the flask is mounted on a frame bridging this area. The frame, the so called bridge, also contains a contamination protection, a door to close the lifting hole as well as sticks for positioning the exchange flask precisely. Inside the flask, a hook couples to the target insert, which pulls it out afterwards. During the beam period of HIPA, the spare target insert is taken out of a shielded parking position. The exchange of the bearings is done in the shielded service cell ATEC at PSI, which is equipped by manipulators and cameras. The exchange of the target insert takes 1.5 to 2 days usually. If possible, the target exchange is moved to a service period, however, the failure of the bearings is difficult to predict. For this, the current needed for the motor to turn the wheel is monitored. Stronger spikes and jumps often indicate that wear debris is blocking the balls in the bearing; however, sometimes it disappears after few days. In this case, the debris could be milled to dust or fell out of the bearing.

3 Recent Improvements

The bearings from the company GMN (Germany), which consist of ceramic balls (Si_3N_4) in a stainless steel cage and rings, the latter coated with MoS_2 and Ag as lubricants, had to be usually exchanged once in the beam period of HIPA, i.e. they last 4 to 5 months. Since this is not optimal, several types of bearings were tested, first in three test stands without proton beam. One type called full ceramics from Cerobear (Germany) had both balls and rings from Si_3N_4, i.e. only the cage was out of stainless steel. Under real conditions on the target in the proton beam it failed after a few weeks. The metal cage was strongly twisted in such a way that some balls fell out of the bearing during the exchange in the service cell. The reason could be that the cage got too hot because of the low conductivity of the ceramic ring, which dissipated only a small amount of heat to the axis. Another bearing ordered from the company Koyo, Germany (belonging to JTEKT, Japan), had still a good performance after 420 days in the test stand [9]. This ball bearing is entirely made out of stainless steel with blocks from WS_2 in between the balls as lubricant. It is also used in the meson production target at J-PARC with a predicted life time of 22 years [10]. During the HIPA shutdown, before the start of the beam period in April 2021, three Koyo bearings were mounted. No change of the bearings was needed. Moreover, the motor current monitored was almost constant compared to previous experience. No degradation of the bearings at the end of the year was observed. It is very likely that this set of bearings could have been operated longer. To avoid the risk of a failure, a new set of Koyo bearings were used, which also lasted the whole beam period in 2022.

Since 2019, several new target types were tested in the Target E station. These developments had two goals: to use the target itself as beam position monitor and to increase the surface muon rate. In the beginning, these developments were followed up independently, however, finally combined. In the first case, the basic idea is to modulate the beam transmission according to the beam position on target. In the case of the slab type target, where the beam passes through the full rim, this can be performed by additional grooves on the left and right side of the rim. If the beam leaves its nominal position in the region of the target centre, part of beam will pass through the grooves on the outer surface of the target. Since the grooves reduce the effective thickness of the target, this leads to a higher beam transmission. As the grooves are placed in a regular but different pattern for the left and right sides of the target, the amplitude of the modulation extracted by Fast Frequency Analysis (FFT) contains information about the beam position. This is much more sensitive than the usual method of measuring the total transmission through the target by comparing the current behind and before the target. Already a beam deviation of 1 mm from the centre increases the modulation by two [11].

To increase the rate of surface muons, the active target surface has to be enlarged. For this, a slanted target geometry has been developed, where the beam impinges on the target under a small angle keeping the effective thickness the same but increasing the surface [9]. Such a target was first tested at the end of 2019. Up to 50% more surface muons were measured and its dependence on the secondary beam lines was confirmed as predicted by simulations. Due to the larger rim of 85 mm instead of 40 mm as well as the slightly smaller width of 5.6 mm, the target is more prone to deformations. In addition, an important difference to the slab target is that the beam heats just the centre part of the rim whereas the surface temperature on the rim in case of the slab target is nearly homogeneous. Indeed, ANSYS calculations showed that the von Mises stresses are larger but do not exceed the ultimate tensile stress limit. No deformation is observed so far, however, the irradiation time is still too short for a final judgement.

Both functions, monitoring the beam position and increasing the surface muon rate, were combined in the newest target development, a slanted type with grooves and ribs. The ribs shown in Fig. 2 vary in thickness and are located on the right of the nominal beam entering the target. On the backside of the target, there are two more rows of ribs and grooves for the left and centre position of the exiting beam. First operation of the target in beam was in 2022, analysis of its performance is underway.

Fig. 2. Slanted graphite target with ribs on the outer surface of the rim (= beam entry). Spokes and hub are mounted.

4 Target M station

The present target M station dates back to 1985 with a renewal of the target insert in 2012. Except for the fact that it is a graphite target rotating with 1 Hz, its design is quite different from the Target E station. Since the target M plug is inserted horizontally, only one long rotational axis is needed and the bearings are shielded. Therefore, the bearings last several years. Since the effective thickness is 5 mm only, the temperature reaches about 1100 K cooled mainly by thermal conductivity. The two beam lines for secondary particles are positioned in the forward direction under 22.5°, i.e. optimal for high energetic particles like pions. However, both users of the beam lines, one used for particles physics, the other for material research, are aiming for low energetic surface muons. Therefore, this target station is subject of renewal in the framework of the HIMB initiative. HIMB belongs together with TATTOOS, the Targeted Alpha Tumor Therapy and Other Oncological Solutions, to the IMPACT application for the Swiss Roadmap of Research Infrastructures from PSI, University of Zurich (UZH) and University hospital Zurich (USZ) and aims to keep PSI at the forefront of research in material, particle physics and life science. Details can be found in the Conceptual Design Report (CDR) [12].

5 HIMB

HIMB aims to increase the muon surface rate by a factor of 100 to 10^{10} μ/s by optimising target and secondary beam lines for this purpose. However, first the Target M station has to be dismantled by remote handling since the residual dose rate, after removing collimators and

target with the exchange flask, is estimated in the order of 2 Sv/h. The key components of the upgrade are:
• A capture solenoid with a large diameter and magnetic field (~0.45 T) in close distance (250 mm) to and on both sides of the target.
• Large transmission for low energetic muons in a series of solenoids with large aperture.
• Increase target thickness from 5 to 20 mm.
• Slanted target type.
The close distance of the capture solenoid to the target requires the compensation of the fringing field at the location of the proton beam, which is otherwise strongly bent upwards. The fringing field is reduced by a mirror plate at the capture solenoid, which has to be cooled due to its close vicinity to the target. Further, an additional steering magnet for the beam is necessary. Due to the four times thicker target, a new collimator system as well as more shielding is needed to cope with the larger losses and radiation. The new target H insert resembles the target E station; however, due to the small amount of space available, it is not possible that the beam passes through the target at the height of the rotational axis, because otherwise it would hit the hub. To save even more space the beam has to pass through the cooling plate behind the graphite wheel (s. Fig. 3). This requires measures to avoid that the beam hits the inside of the copper plate. For this, an aperture loss monitor is mounted in front of a current sensitive collimator out of tungsten in the cooling plate. Both can trigger the interlock system to switch off the beam.

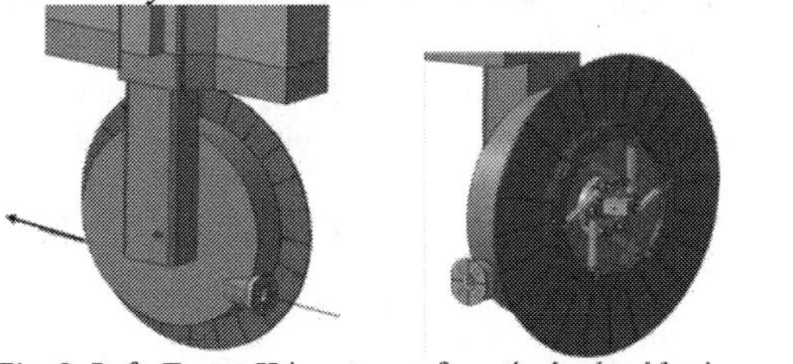

Fig. 3. Left: Target H insert seen from the back with view on the cooling plate. The arrow indicates the proton beam. Right: View on the graphite wheel with mounted spokes and hub.

The slanted target (Fig. 3) is the favourite version. It is flatter compared to target E and consists of two disks, each containing every second segment with 3.5 mm thickness. ANSYS simulations look promising regarding the expected deformation and thermal stress.

6 Summary

The operation of a meson production and muon target at 50 kW poses several challenges like cooling, thermal stress and high radiation. With the Koyo bearings from J-PARC an important step forward was achieved to increase the reliability of Target E. This is particularly important for the new Target H station, foreseen in the IMPACT initiative as upgrade of Target M to a 100 times larger surface muon rate. It is planned for 2027 with first beam after the HIPA shutdown in 2028. Target H will also profit from the operational experience with the Target E station as well as from the recent target developments, particularly the slanted target type, which was proven to increase the surface muon rate up to 50%.

Acknowledgement

We would like to thank all, who are contributing to or supporting the meson production targets and IMPACT. The large progress we made would not be possible without the ambition, dedication and expert knowledge of so many people.

We appreciate very much the financial and general support of the PSI directorate and the four divisions NUM, GFA, NES and BIO as well as the strong support of the service groups for infrastructures in LOG.

References

1. J. Grillenberger, C. Baumgarten, M. Seidel, SciPost Phys. Proc., vol. **5**, 002, 2021. doi:10.21468/SciPostPhysProc.5.002
2. D. Kiselev et al., SciPost Phys. Proc. **5**, 003 (2021). doi:10.21468/SciPostPhysProc.5.003
3. F. Berg et al., Phys. Rev. Accel. Beams **19**, 024701 (2016). doi:10.1103/PhysRevAccelBeams.19.024701
4. G. Heidenreich, *Carbon and beryllium targets at PSI*, AIP Conf. Proc. **642**, 122 (2002). doi:10.1063/1.1522602
5. A.D. Hillier et al., *Muon spin spectroscopy*, Nature Review Methods Primer 2, 4 (2022). https://doi.org/10.1038/s43586-021-00089-0
6. N. Maruyama, M. Harayama, J. Nucl. Mat. **195**, 44 (1992). https://doi.org/10.1016/0022-3115(92)90362-O
7. S. Heinitz et al., Applied Radiation and Isotopes **130**, 260 (2017).
8. D. Kiselev et al., J. Radioanal. Nucl. Chem **305**, 769 (2015).
9. D. Kiselev, et al., *Progress and Challenges of the PSI Meson Targets and Relevant Systems*, in Proceedings of the 3rd J-PARC Symposium, J-PARC2019, JPS Conf. Proc. **33**, 011102 (2021). doi:10.7566/JPSCP.33.011102
10. S. Makimura et al., *Perspective of Muon Production Target at J-PARC MLF MUSE*, in Proceedings of the 14th Int. Conf. on Muon Spin Rotation, Relaxation and Resonance, μSR2017, JPS Conf. Proc. **21**, 011058 (2018). doi:10.7566/JPSCP.21.011058
11. P. Duperrex et al., *New Centering Beam Monitor for High Power Proton Beam Rotating Target*, in Proceedings of the 22nd Int. Conf. on Cyclotrons and their Applications, CYC2019, Cape Town, South Africa, Sept. 2019, p. 189 (2019). doi:10.18429/JACoW-Cyclotrons2019-TUP016
12. R. Eichler, D. Kiselev, A. Koschik, A. Knecht, N. van der Meulen, R. Scheibl (editors), *IMPACT Conceptual Design Report*, PSI report **22-01** (2022). https://www.psi.ch/en/media/71845/download

Design, testing, commissioning, and early operation of the third-generation n_TOF neutron spallation target at CERN

Raffaele Esposito[1*], *Marco* Calviani[1], and *Oliver* Aberle[1]

[1]European Laboratory for Particle Physics (CERN), 1211 Geneva 23, Switzerland

Abstract. The n_TOF facility at the European Laboratory for Particle Physics (CERN) is a top-class, high-brightness neutron spallation source dedicated to high-resolution neutron time-of-flight experiments. During CERN's Long Shutdown 2 (LS2, 2019-2021), the facility's neutron spallation target was upgraded and is now operating with its third-generation target. The target is based on a pure-Pb core cooled by gaseous nitrogen and has been designed to withstand the impact of a 20-GeV/*c* proton beam in bunches of 10^{13} protons with a bunch duration of 6-8 ns RMS. The produced neutrons span 11 orders of magnitude in kinetic energy (from thermal to GeV) and are delivered to two experimental areas and a neutron irradiation station near the target. This contribution includes a description of the physics and engineering design processes that brought the facility from its second-generation target to its current third-generation target, the planned autopsy of the second-generation target in 2023 to investigate its status after 10 years of operation, tests under beam irradiation carried out on target prototypes at CERN's HiRadMat facility, and the commissioning with beam of the new target, as well as the challenges encountered during the first year of its operation.

1 Introduction on the n_TOF facility

The neutron time-of-flight (n_TOF) facility operates at the European Laboratory for Particle Physics (CERN) in Geneva, Switzerland. It is composed of a pure-lead based neutron spallation target coupled to two neutron flight paths and two experimental areas for neutron time of flight experiments [1]. There is also a neutron irradiation and measurement station located close to the target [2, 3].

The CERN's Proton Synchrotron (PS) delivers 20-GeV/*c* protons to the target, producing neutrons through the nuclear spallation process. These neutrons travel towards two experimental areas: Experimental Area 1 (EAR1) located 185 m behind the target, and Experimental Area 2 (EAR2) located 20 m above the target (Fig. 1).

The PS is currently able to deliver short bunches with a maximum of 8.5×10^{12} protons in 6-8 ns (RMS) to the target, with a minimum period between bunches of 1.2 s. The number of neutrons reaching the experimental areas is on the order of 10^6 n/cm^2 per bunch in EAR1 and 10^8 n/cm^2 per bunch in EAR2.

The neutrons generated by the spallation process are moderated by two water layers: 4 cm of borated water (1.28 wt% of ^{10}B) for the neutrons directed towards EAR1, and 3.4 cm of demineralized water for the neutrons directed towards EAR2.

Thanks to the neutron moderation, a wide neutron energy spectrum is available for the two experimental areas, ranging from thermal energy up to 1 GeV. Moreover, the long flight paths provide high energy resolution ($\Delta E/E = 10^{-4}$ in EAR1 [4, 5]). These features make the n_TOF facility ideal for high signal-to-background ratio measurements of neutron cross-sections of radioactive isotopes, even if they are available in tiny amounts.

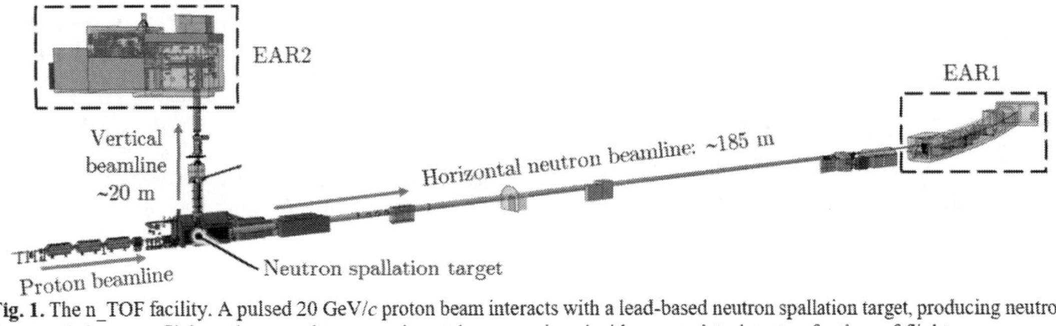

Fig. 1. The n_TOF facility. A pulsed 20 GeV/*c* proton beam interacts with a lead-based neutron spallation target, producing neutrons that travel along two flight paths toward two experimental areas equipped with appropriate detectors for time-of-flight measurements.

* Corresponding author: raffaele.esposito@cern.ch

2 The third-generation n_TOF target

The final design of the third-generation target (Fig. 2) consists of a series of pure-lead (>99.9 wt%) plates cooled by nitrogen gas [6] at slightly above atmospheric pressure. The use of a gaseous coolant instead of a liquid one was motivated by corrosion and erosion issues encountered during the operation of the first and second-generation targets, which were water-cooled in direct contact with lead. Since pure lead is highly susceptible to creep deformation, the lead plates are supported by an anti-creep structure made from aluminum alloy 6082-T6. The target core is contained in a stainless-steel vessel, and the water moderator vessels are made from aluminum alloy 5083. The EAR1 aluminum moderator vessel is directly bonded to the stainless-steel vessel via explosion welding, with intermediate layers of pure aluminum and titanium between stainless steel (vessel side) and aluminum 5083 (moderator side). The target was originally designed with the beam parameters from Table 1 as a reference.

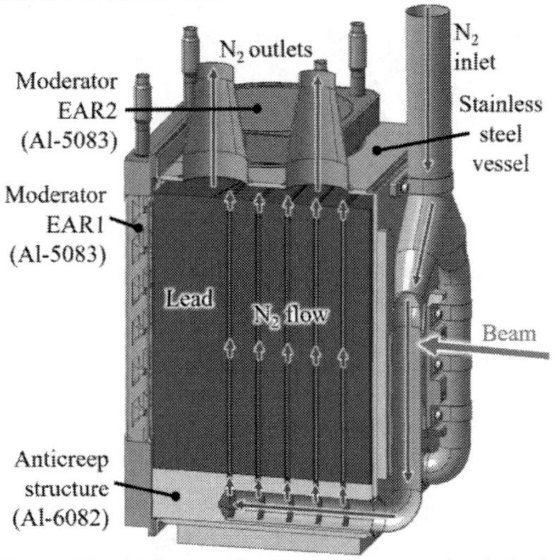

Fig. 2. The third-generation n_TOF target (sectional view). Six pure-lead (>99.9 wt%) plates are supported by an aluminum 6082-T6 anti-creep structure. Pure nitrogen gas flows in the channels between the lead plates to cool them. The target core is enclosed in a stainless-steel vessel. The two moderator vessels are made from aluminum 5083, and the EAR1 moderator vessel is directly bonded to the stainless-steel vessel via explosion welding.

Table 1. Beam parameters used as a reference in the design of the third-generation n_TOF target.

Bunch intensity	10^{13} p^+
Beam momentum	20 GeV/*c*
Bunch energy	32 kJ
Bunch length	25 ns (4σ)
Beam size (Gaussian)	15 mm (1σ)
Average intensity	1.67×10^{12} p^+/s
Average current	0.27 μA
Average power	5.4 kW
Peak current	91.3 A
Peak power	1.8 TW

3 Thermo-mechanical behavior

The third-generation neutron target has been in operation since July 2021. The design studies for this new target began in 2015 taking into account the operational experience of previous spallation targets and analysis of different target concepts based on materials such as lead, tantalum, and tungsten, as well as cladding layers of tantalum, titanium Grade 5, and Inconel. Various cooling fluids were also compared, including water, air, nitrogen, and argon. A more detailed summary of the design solutions explored and the rationale behind the final target design can be found in Ref. [7].

To evaluate the performance of each design solution, Monte Carlo simulations (primarily using the FLUKA code [8, 9]) were performed to estimate the physics performance (e.g., neutron flux, energy resolution, photon background) and the heat distribution induced in the target by interaction with the proton beam. Computational Fluid Dynamics (CFD) simulations were used to estimate the heat removed per unit of surface area and temperature gradient (Heat Transfer Coefficient, HTC) by the target cooling system. The heat distribution generated by the beam - target interaction and the HTC map on the target surfaces were used as boundary conditions for Finite-Element (FE) thermal analyses, which provided the time evolution of the temperature distribution in the examined target solutions. This resulting temperature distribution was then used as boundary condition for Finite-Element structural analyses to estimate the time evolution of the stress distribution in the target during operation.

The results of these studies showed that each instantaneous temperature rise caused by a beam impact leads to a rapid thermal expansion of the target core followed by the propagation of plastic stress waves inside the target. The temperature distribution in the target and the resulting stresses from stress-wave propagation are the main factors to consider when assessing the thermo-mechanical integrity and fatigue resistance of the target [10].

In the final design, and with the beam parameters listed in Table 1, thermal simulations (see Fig. 3) estimate a peak temperature of 135°C inside the lead plates in the worst-case scenario (six consecutive bunches impacting the target, separated by 1.2 s and followed by a cooling period of 30 s). Structural finite-element simulations revealed that the peak von Mises stress in the pure lead plates is 2.6 MPa, indicating that the material operates within the elasto-plastic regime. As a result, the material may be susceptible to cyclic plasticity and fatigue damage during the target's lifetime, which spans millions of cycles (10 million cycles were considered for fatigue calculations). To validate the target's ability to perform its intended function over its entire lifespan, a test of a prototype of the target under beam irradiation was conducted at the HiRadMat facility at CERN. The test is described in Section 5.

4 Modeling of creep behavior

Creep deformations of pure lead represent a critical aspect of the target's thermo-mechanical behavior as pure lead could deform into the cooling channels, partially blocking them and compromising the cooling efficiency. A dedicated material constitutive model was implemented in the Finite-Element codes to simulate the creep behavior of lead under the different conditions of stress and temperature that can be found in the target during operation. The model is based on creep data for lead provided in Ref. [11]. It consists of a series of piecewise functions, each composed of multiple power laws (Fig. 4):

$$\dot{\varepsilon} = c\,\sigma^n \quad (1)$$

where $\dot{\varepsilon}$ is the plastic strain-rate due to creep, σ is the von Mises equivalent stress, c and n are coefficients whose values best fit the available data. The coefficient n is close to 1 when the main mechanism of creep flow is diffusional flow, while it can reach values between 5 and 7 in the case of power-law creep. Each curve in Fig. 4 is valid for a specific temperature; values for intermediate temperatures are obtained by logarithmic interpolation.

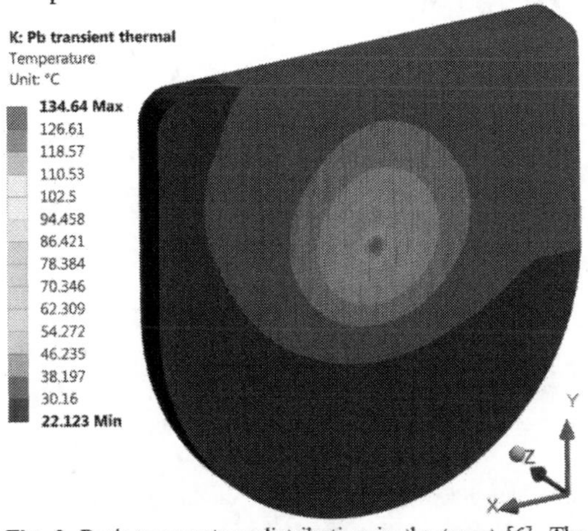

Fig. 3. Peak temperature distribution in the target [6]. The image shows the model of the second lead plate, where a peak temperature of 135°C is reached. The parameters of the beam load are listed in Table 1.

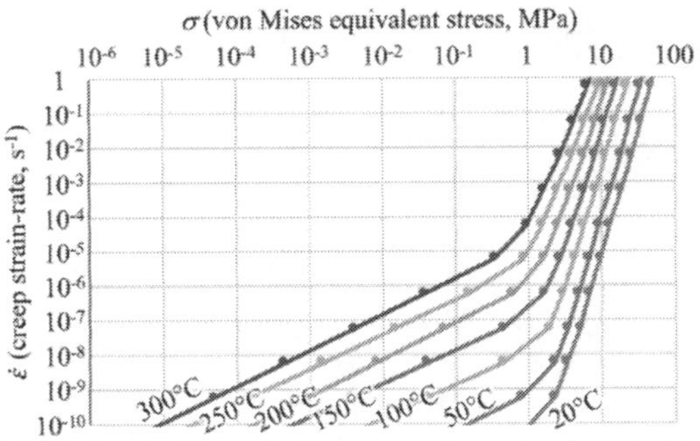

Fig. 4. Constitutive model for pure-lead creep implemented in Finite-Element code. It consists of a series of power laws. Curves for intermediate temperatures are obtained by logarithmic interpolation. Data from Ref. [11].

The proposed creep model has been implemented in Finite-Element codes to simulate the creep deformations in the target after 6 years and 4 months of continuous operation (twice the expected target's lifetime) under a steady peak temperature field. The maximum lead penetration into a cooling channel was found to be 0.64 mm, whereas the depth of the cooling channels is 3 mm (see Fig. 5). Accident scenarios, including the possibility of fully obstructed channels, were considered in CFD simulations. The results of this analysis showed that the target's operation would not be compromised: the new peak temperature would be equal to 160°C, which is considered acceptable for the accident scenario examined.

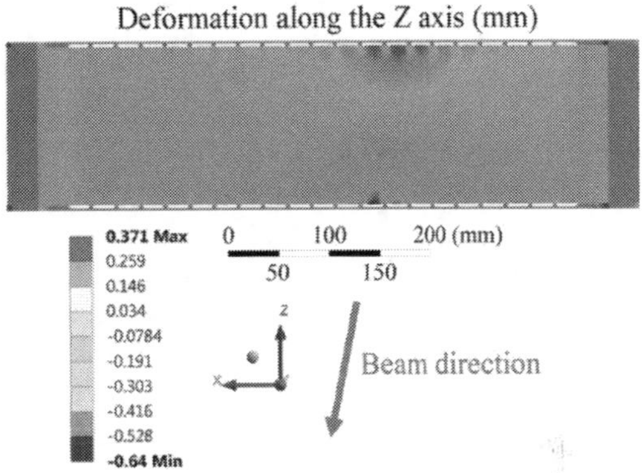

Fig. 5. Simulation of creep deformation in the lead target [6]. Sectional view from above. The maximum penetration of lead into a cooling channel is 0.64 mm, in the sixth lead plate.

5 Tests under beam irradiation

A small-scale prototype of the third-generation target was tested at the HiRadMat facility at CERN. The prototype consisted of six pure-lead blocks enclosed in an aluminum vessel filled with nitrogen gas. It was irradiated with 440-GeV/*c* proton-beam pulses from the CERN's Super Proton Synchrotron (SPS), with a maximum of 3.5×10^{13} protons delivered in short pulses of 8 μs [12].

The prototype was hit by 1500 proton-beam bunches of 4–4.5×10^{12} protons with a period of 22.8 s and a beam size of 4 mm RMS. These parameters were chosen to induce in the prototype a slightly higher fatigue damage than the one induced in the n_TOF third-generation target under the nominal beam loads described in Table 1. The temperature of the lead blocks was monitored by thermocouples and used to automatically control heating foils that were glued to the aluminum container to maintain the desired temperature range of 100–140°C. This temperature range is the one estimated by the thermal simulations described in Section 3 and Fig. 3.

The pure-lead blocks were inspected by neutron tomography at the Paul Scherrer Institut (PSI) with an image resolution of 100 μs. No defects were detected inside the material, confirming the soundness of the target design and its ability to produce neutrons for the facility for another decade.

6 Commissioning with beam

The new target began operating in July 2021. The profile of the proton beam is monitored using a Secondary Emission Monitor (SEM) grid. According to the design assumptions in Table 1, a beam size of 15 mm was expected. However, during the commissioning with beam of the new target, the beam size appeared to be too large for the vertical aperture of one of the bending dipoles upstream of the target. This led to beam losses and a high dose rate in the area. The beamline was temporarily operated with a narrower beam on the vertical axis (7 mm RMS), which was not acceptable for the long-term integrity of the target. The issue was solved by removing the last two focusing quadrupoles upstream of the target from the proton beamline during the 2021-2022 winter CERN's accelerator stop. This enabled the optimization of the beam optics and the possibility to deliver a larger beam spot on the target with contained beam losses. As a result, the beam spot measured at the SEM grid had a size of 27 mm along the horizontal axis and 15 mm along the vertical axis. This beam size is now even larger than the one in the design assumptions of Table 1, enabling an increase in the average intensity on target up to 2.3×10^{12} p^+/s (7.4 kW).

7 Conclusions

The third-generation n_TOF target has been successfully operational since July 2021. Thermo-structural simulations were used to simulate the target's response when impacted by high-intensity beam pulses. These simulations revealed that the temperatures and stresses reached during operation bring the pure lead in the target into elasto-plastic regime, raising concerns about cyclic plasticity and fatigue damage. A test was designed and performed with a prototype under beam irradiation to validate the target's capability to maintain its integrity under repeated beam impacts, which was successfully validated. Simulations were exploited to design the test in a way that the estimated fatigue damage in the prototype would reproduce that of the target operating under nominal conditions, with some margin for conservatism. Additionally, a new material constitutive model was implemented in the FE software to estimate creep deformations. This was done because of concerns that creep deformation could obstruct cooling channels and impair cooling efficiency, but simulations showed that the deformations were acceptable and the obstruction was contained within acceptable limits. Finally, similar simulations were employed to estimate the consequences of accident scenarios and confirmed that there would be no catastrophic consequences.

A better understanding of the plastic deformations induced in the target by creep and beam impacts will be achieved by examining the second-generation target, also based on pure lead. This target operated from 2008 to 2018 [7] and has a different geometry, with the core made of a single massive lead cylinder with a diameter of 60 cm and a length of 40 cm; it was cooled by water rather than nitrogen gas. Despite the different geometry, the examination of the lead surfaces of the spent target can contribute to the validation of the estimations of the plastic deformations due to creep and beam impacts on the target. The lead target core is enclosed in an aluminum vessel, which includes the target's borated-water moderator vessel. To expose the surfaces of the lead core, the aluminum vessel must be machined. One base of the lead cylinder will be exposed by machining the side of the vessel directly impacted by the proton beam, while the other base of the cylinder will be exposed by machining the moderator vessel. Due to the high radioactive dose around the target, the procedure will be executed by robots. The removed moderator vessel will be inspected to assess the presence of deposited boron from the borated water and pitting corrosion effects. The exposed lead surfaces will then be scanned by a 3D laser scanner, providing the lead surface deformations due to creep and beam impacts after 10 years of target operation.

The authors would like to acknowledge the support of the n_TOF collaboration, the n_TOF third-generation target team at CERN, the workgroup of the second generation target autopsy & waste packaging, the HiRadMat team, the NEUTRA facility at the Paul Scherrer Institut for the neutron tomography of the irradiated target prototype, and the FTN line workgroup for the effort to deliver an optimal proton beam to the target.

References

1. C. Guerrero et al., Eur. Phys. J. A **49**, 27 (2013)
2. M. Ferrari, D. Senajova et al., Phys. Rev. Accel. Beams **25**, 103001 (2022)
3. N. Patronis et al., arXiv:2209.04443 (2022)
4. M. Barbagallo, C. Guerrero, A. Tsinganis, D. Tarrío et al., Eur. Phys. J. A **49**, 156 (2013)
5. M. Sabaté-Gilarte, M. Barbagallo, N. Colonna et al., Eur. Phys. J. A, **53**, 210 (2017)
6. R. Esposito et al. (n_TOF Collaboration), Phys. Rev. Acc. Beams **24**, 093001 (2021)
7. R. Esposito, M. Calviani, J. Neutron Res. **22**, 221 (2020)
8. T.T. Böhlen et al., Nucl. Data Sheets **120**, 211 (2014)
9. A. Ferrari, P.R. Sala, A. Fassò, J. Ranft, FLUKA: a multi-particle transport code (program version 2005) (CERN, Geneva, 2005)
10. R. Esposito, *Design, prototyping, and thermos-mechanical modelling of a neutron spallation target impacted by high-energy proton-beam pulses in the n_TOF facility at CERN*, Ph.D. dissertation, EPFL, Lausanne (2022)
11. H.J. Frost, F. Ashby, Deformation-mechanism maps: the plasticity and creep of metals and ceramics (Pergamon Press, Oxford, 1982)
12. I. Efthymiopoulos et al., *HiRadMat: a new irradiation facility for material testing at CERN*, in Proceedings of the International Particle Accelerator Conference, IPAC, 4-9 Sep. 2011, San Sebastián, Spain (2011)

A spallation target at TRIUMF for fundamental neutron physics

Sean Vanbergen[1,2*], *Wolfgang* Schreyer[1], *Shomi* Ahmed[3], *Taraneh* Andalib[3], *Mike* Barnes[4], *Beryl* Bell[1,5], *Jason* Chak[1], *Elspeth* Cudmore[1,6], *Beatrice* Franke[1,2], *Pietro* Giampa[1], *Patricia* Gnyp[1,7], *Moritz* Hahn[7], *Sean* Hansen-Romu[3], *Kichiji* Hatanaka[8], *David* Jones[2], *Katerina* Katsika[1], *Shinsuke* Kawasaki[9], *Elie* Korkmaz[10], *Kirk* Madison[2], *Juliette* Mammei[3], *Russell* Mammei[1,3,11], *Cameron* Marshall[1], *Jeffery* Martin[1,11], *Ryohei* Matsumiya[1,8], *Kenji* Mishima[9,12], *Takahiro* Okamura[9], *Ruediger* Picker[1], *Edgard* Pierre[1], *Des* Ramsay[1], *Gelo* Remon[1], *Steve* Sidhu[1,13], *Brendan* Thorsteinson[11], and *Willem* van Oers[1]

[1]TRIUMF, Vancouver, BC, Canada
[2]University of British Columbia, Vancouver, BC, Canada
[3]University of Manitoba, Winnipeg, MB, Canada
[4]Center for European Nuclear Research, CH-1211 Geneva 23, Switzerland
[5]McGill University, Montreal, QC, Canada
[6]Carleton University, Ottawa, ON, Canada
[7]Coburg University of Applied Science, Coburg, Germany
[8]Research Center for Nuclear Physics, Osaka University, Ibaraki, Osaka, Japan
[9]High Energy Accelerator Research Organization, KEK, Tsukuba, Ibaraki, Japan
[10]University of Northern British Columbia, Prince George, BC, Canada
[11]University of Winnipeg, Winnipeg, MB, Canada
[12]J-PARC, Tokai, Ibaraki, Japan
[13]Simon Fraser University, Burnaby, BC, Canada

Abstract. Ultracold neutrons (UCNs) are a powerful tool for probing the Standard Model at high precision. The TRIUMF Ultracold Advanced Neutron (TUCAN) collaboration is building a new UCN source to provide unprecedented densities of UCNs for experiments. This source will use a tantalum-clad tungsten spallation target, receiving up to 40 μA of 480-MeV protons from TRIUMF's main cyclotron. The beamline and target were constructed from 2014 to 2016 and operated at beam currents up to 10 μA from 2017 to 2019 as part of a prototype UCN source. We describe the design choices for the target and target-handling system, as well as our benchmarking of the target performance using UCN production measurements.

1 Introduction

Ultracold neutrons (UCNs) are neutrons with kinetic energies less than the optical Fermi potential of certain solid materials, with the upper cut-off corresponding to the potential of Ni-58 at 347 neV. Fulfilling this energy condition allows UCNs to be reflected from material surfaces under any angle of incidence, and so they can be contained and studied for hundreds of seconds.

This containment property has made UCNs a powerful tool for probing fundamental physics and testing extensions of the standard model. Measurements of the neutron lifetime and beta decay asymmetry test the weak interaction, including unitarity of the CKM matrix [1]. Searches for electric dipole moments [2] are a precise test of CP violation and can help explain the observed matter-antimatter asymmetry in the universe [3, 4]. UCNs also provide a rare system where gravity can be observed in a quantum-mechanical context and allow non-Newtonian gravity to be tested [5, 6].

Improving the precision of such experiments requires achieving higher UCN densities in experimental apparatuses. Modern UCN sources are based on the conversion of cold neutrons to UCNs through interactions with fundamental excitations (phonons and rotons) in bulk materials. This process is non-conservative and can therefore produce high phase-space densities, enabling similarly high real-space densities in experiments. UCN sources utilizing this technique have been developed using solid D_2 [8-11] and superfluid He-II [12-14].

At TRIUMF, the TUCAN collaboration is building a new UCN source, using room-temperature D_2O and 20 K D_2 neutron moderators to produce cold neutrons and He-II as a UCN converter [15]. This source will achieve a world-class UCN density and enable a new generation of high-precision neutron experiments [16].

Neutron production in the TUCAN source is achieved by spallation. A new 480-MeV proton beamline, driven by the TRIUMF main cyclotron, has been constructed for this purpose, which we have described in previous publications [17]. In these proceedings, we describe the details of the target design, handling, and analysis of its performance.

* Corresponding author: svanbergen@triumf.ca

2 Target design

The TUCAN source target, shown in Fig. 1, is a set of five tantalum-clad water-cooled tungsten blocks, enclosed in a stainless-steel flask through which the cooling water circulates.

Tungsten has long been considered an ideal material for the construction of solid spallation targets due to its high neutron yield and high strength at elevated temperatures. The choice of tungsten as a target material was driven by the successful design of such a target for the now-decommissioned KENS facility [18]. Targets of a similar composition have continued to operate with great success at LANSCE [19] and ISIS [20]. Tungsten targets will also be used the ESS, although with a substantially different design [21].

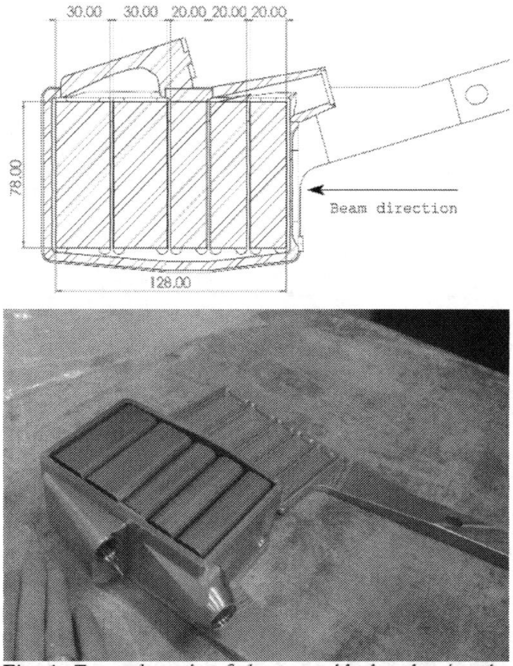

Fig. 1. Top: schematic of the target blocks, showing the dimensions of the blocks in mm. The height of the blocks, not shown, is 56 mm. Bottom: photograph of the target flask with the blocks inside. The inlet (front) and outlet (rear) of the flask are visible in the bottom left corner.

The limitation of tungsten as a target material is that it experiences a high rate of corrosion in water under high temperature or radiation. The solution to this has been to clad tungsten targets with a second material that is resistant to corrosion [18]. Tantalum has been studied as an ideal material for this cladding, as it also has a high neutron yield. A 0.86-mm-thick tantalum cladding was applied to the TUCAN target blocks using hot isostatic pressing (HIP). This process was performed by Allied Materials Inc. at a temperature of 1500 °C and a pressure of 200 MPa in high-purity (5N) argon.

At the maximum current of 40 μA and a typical proton energy of 480 MeV, the beam will deliver approximately 20 kW to the target. Simulations of the spallation process using MCNP [22] indicate that 70% of the energy is deposited as heating of the target, with the remaining 30% being carried away by the spallation products. The energy deposited in the target is removed by a convective water-cooling loop.

The target geometry has been optimized alongside the water-cooling to reduce heating and stress. A total target thickness of 12 cm was chosen such that all protons are stopped in the target. A single 12-cm-thick block would experience un-manageable heating, and so it is necessary to split the target into several blocks.

Extensive finite element simulations of the target heating and water cooling were carried out using ANSYS to determine a suitable geometry, an example of which is shown in Fig. 2. These studies explored different numbers of blocks and different block thicknesses to arrive at the chosen geometry, shown in figure 1. This geometry consists of five blocks, with the first three being 20 mm thick, and the last two being 30 mm thick. The blocks are separated by 2 mm wide channels through which the cooling water circulates with a total flow rate of 0.75-0.9 kg/s at 70 psig supply pressure.

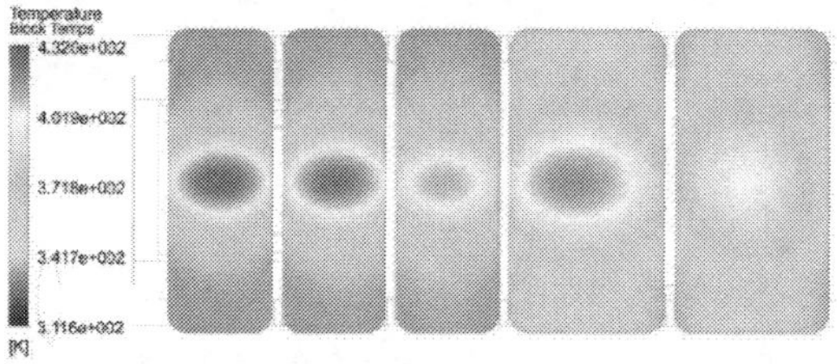

Fig. 2. Finite-element simulated temperature profiles in the target blocks from heating by the 40-μA 480-MeV proton beam. The profiles are cut through the centre of the blocks, corresponding to the centre of the beam where maximum heating occurs.

Table 1. Maximum stresses and temperatures achieved in different parts of the target.

Location	Maximum Stress (MPa)	Maximum Temperature (°C)
Target blocks	96	158
Flask window	88	72
Cooling water	N/A	119

The resulting maximum temperatures and stresses in the target are shown in Table. 1. The steel of the flask has a yield strength of 205 MPa. The cooling water should be kept below its boiling point, which at the pressure of the water supply is 158 °C. Both conditions are satisfied by the design.

Design parameters for the target were originally chosen based on the experience of the KENS target, which indicated that the temperature gradient in each block should be kept below 200 °C, and the stress below 200 MPa [23].

Due to the higher beam current compared to KENS (which was operated up to 10 μA), the target may see faster loss of strength due to irradiation damage. From FLUKA [24,25] simulations, the target is expected to see a maximum 1.1 dpa and 118 He appm per year. Recent studies have characterized the effect of such levels of irradiation damage on tungsten and tantalum [26]. Based on these results, the target's ultimate strength should remain above 200 MPa after 10 years of operation, after which the target will be replaced.

3 Target handling system

The target is inserted into a helium-filled aluminium crypt at the end of the proton beamline. The crypt is separated from the vacuum beam pipe by a 1-mm-thick Al-2219 window. This material was chosen due to its superior strength at high temperature, as simulations of the window indicate that it could reach up to 120°C during operation. The purpose of the crypt is to provide a separating region between the target and the rest of the beamline, especially in the event of any damage to the target such as a cooling-water leak. The helium atmosphere provides some additional cooling of the crypt while not being significantly activated by the beam.

The target is held in the crypt by the target arm, shown in Fig. 3. The arm is built in three sections: the fore-arm, the mid-arm, and the carriage. The fore-arm directly connects to the target flask, and is connected to the mid-arm with a special disconnection joint. This joint allows the fore-arm, with the target flask, to be disconnected for disposal. The mid-arm connects the fore-arm and the carriage. The carriage is a large steel block that provides a counterweight to support the cantilever of the arm and serves as shielding against prompt radiation from the target. It also serves as part of the target extraction system.

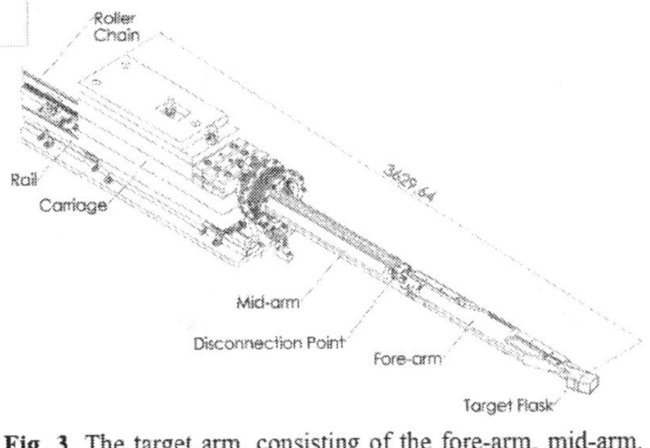

Fig. 3. The target arm, consisting of the fore-arm, mid-arm, and carriage, which together support the target in the beamline. The carriage rests on a pair of rails, and is connected via a roller chain to a gearbox (not shown) which is used to move the target arm into and out of the beamline.

The extraction system is formed by a set of hardened-steel rails on which the carriage rests. At the end of the rails is a gearbox, connected to the carriage by a roller chain. During manoeuvring of the target, the gearbox is driven by a long pole, allowing the target to be moved without exposure to high dose rates.

The target extraction has been designed with a minimum target life of 10 years. After reaching end of life, the target must be removed and stored. At this point, the target will be heavily irradiated and must be placed in long-term shielded storage. For that purpose, a lead-lined storage cask has been built.

The target cask was developed using FLUKA simulations to determine the dose rate. After operating the target for 20 years with an average beam current of 10 μA (a 25% duty cycle on the 40 μA beam current) the equivalent-dose rates close to the target can exceed 1 Sv/h. Extraction of the target requires removal of concrete shielding, which is only possible during the yearly 3-months-long shutdown of the TRIUMF cyclotron. During this time, short-lived radionuclides will be allowed to decay for a cooldown period of approximately one month. Then the target will be extracted and placed in the cask. Fig. 4 shows the equivalent-dose rate at 50 cm from the target cask as a function of cooldown time. After a 28-day cooldown period, the equivalent-dose rate is reduced below 200 μSv/h, allowing the cask to be safely handled by trained personnel and transported to a long-term storage location.

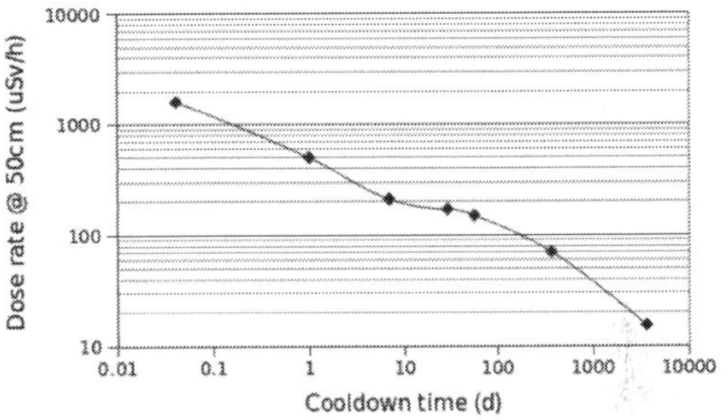

Fig. 4. Equivalent-dose rate, calculated using FLUKA, at 50 cm distance from the target cask, once the target has been placed in the cask, following 20 years of operation and a cooldown period to allow short-lived radionuclides to decay.

4 Performance benchmarking

The performance of the target has been tested by comparing simulation results to measurements of UCN production.

In 2017, we installed a prototype UCN source, which was originally developed and operated at RCNP Japan [27], above the spallation target. This source used He-II as a UCN converter, but with a significantly lower cooling power compared to the new TUCAN source, which limited the operation current. This source used graphite, heavy water, and heavy-water ice cooled to 20 K to moderate the fast spallation neutrons to cold neutrons. The cold neutrons were then converted to UCNs in an 8-litre volume of He-II cooled to 1 K. The geometry of this source relative to the target are described in detail in our previous publication [28].

During 2017, 2018, and 2019 we operated the beamline, spallation target, and UCN source for a month each year [28]. During these experimental campaigns, measurements were made to understand the production of UCN by the source. For such a measurement, we typically irradiated the target for 60 s while a valve confined the UCN inside the source. Then, the valve was opened and the UCN accumulated in the source diffused along stainless-steel guides to a detector.

A fast kicker magnet installed in the beamline [29] allows us to vary the average beam current diverted to the target. Fig. 5 shows the expected linear proportionality between beam current and UCN counted in the detector up to 1 μA. We have operated the beamline at higher currents, up to 10 μA, but due to limited cooling power to maintain the He-II temperature, the prototype UCN source was not able to handle the increased heat loads at higher currents. This causes us to observe a decrease in the UCN yield.

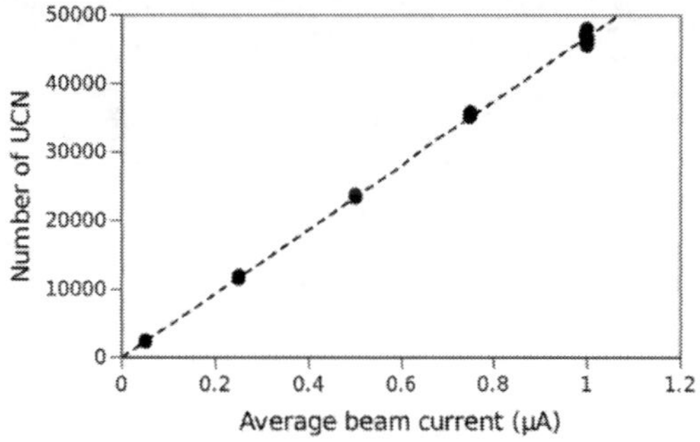

Fig. 5. UCN extracted from the source after irradiating the target with varying average beam currents for 60 s. The UCN yield increases linearly with current up to 1 μA [17, 28]. The larger scatter in the points at 1 μA can be attributed to fluctuations in the source conditions.

To understand the neutron production, we modelled the target, moderators, and UCN source in MCNP [22], shown in fig. 6. The UCN production rate in the prototype source expected from these simulations was 21000 $s^{-1}\mu A^{-1}$. This number was used as an input for UCN Monte Carlo transport simulations using PENTrack [30]. The results of these simulations showed good agreement with the detected number of 42000 UCNs at 1 μA beam current [28].

In this analysis, there were significant uncertainties in the transport properties of UCN guides and in temperature-dependent UCN losses inside the source, which resulted in equally large uncertainties on the exact performance parameters of the source. Despite these limitations, the results are consistent with the target operating as expected.

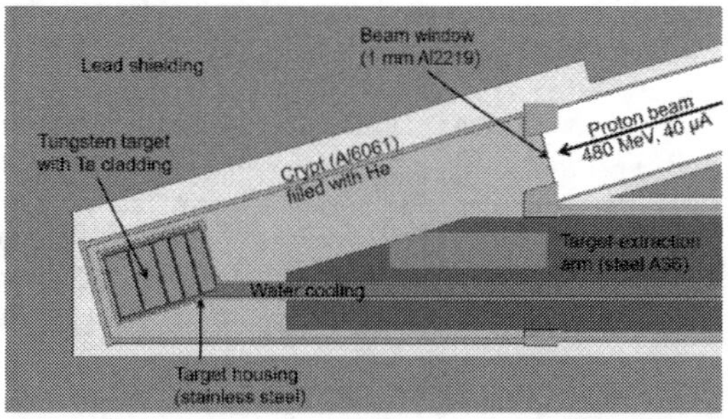

Fig. 6. Simulation model of the target (top view) [27, 31].

5 Conclusion

We have developed and installed a tantalum-clad tungsten spallation target as part of the TUCAN source. This target has been designed to accept 40 μA of 480 MeV proton beam current with an expected operational lifetime of 10-20 years. All target handling systems have been completed and the target is ready for long-term operation. The target performance has been operated at currents up to 10 μA. UCN production data up to 1 μA has been analysed and found to be consistent with the expected source performance.

Beginning in 2024, the target will be operated as part of the new TUCAN source at currents up to 40 μA. Based on our predictions of the source performance [16], we expect the TUCAN source to be capable of producing the highest density of UCN in experiments ever achieved. This will enable a new generation of high-precision experiments using neutrons.

References

1. D. Dubbers and B. Markisch, Annu. Rev. Nucl. Part. 71, 139-163 (2021)
2. J. W. Martin J. Phys.: Conf. Ser. 1643, 012002 (2020)
3. V. Cirigliano, Y. Li, S. Profumo, and M. J. Ramsey-Musolf, J. High Energy Phys. 2010 (2010)
4. S. J. Huber, M. Pospelov, and A. Ritz, Phys. Rev. D 75, 036006 (2007).
5. R. I. Sedmik, J. Bosina, L. Achatz, et. al., EPJ Web Conf. 219, 05004 (2019).
6. B. Clement, S. Baeßler, V. V. Nesvizhevsky, et. al., arXiv:2205.11130 (2022).
7. T. Jenke, G. Cronenberg, J. Burgdorfer, et. al., Phys. Rev. Lett. 112, 151105 (2014)
8. T. M. Ito, E. R. Adamek, N. B. Callahan, et. al., Phys. Rev. C 97, 012501 (2018).
9. J. Kahlenberg, D. Ries, K. U. Ross, et. al., Eur. Phys. A 53 (2017)
10. B. Lauss and B. Blau, SciPost Phys. Proc. 004 (2021).
11. E. Korobkina, G. Medlin, B. Wehring, et. al., NIM A 767, 169 (2014)
12. F. M. Piegsa, M. Fertl, S. N. Ivanov, et. al., Phys. Rev. C 90, 015501 (2014).
13. Y. Masuda, K. Hatanaka, S.-C. Jeong, et. al., Phys. Rev. Lett. 108, 134801 (2012).
14. A. P. Serebrov, A. Fomin, M. S. Onegin, et. al., Tech. Phys. Lett. 40, 10 (2014)
15. J. Martin, B. Franke, K. Hatanaka, S. Kawasaki, and R. Picker, Nucl. Phys. News 31, 19 (2021).
16. S. Sidhu, W. Schreyer, S. Vanbergen, et. al., Eur. Phys.: Web. Conf. (2022) (To be published)
17. S. Ahmed, T. Andalib, M. Barnes et al., Nucl. Instr. Meth. A 927, 101 (2019)
18. M. Kawai, K. Kikuchi, H. Kurishita, et. al., J. Nucl. Mats. 296, 312-320 (2001)
19. A. T. Nelson, J. A. O'Toole, R. A. Valicenti, S. A. Maloy, J. Nucl. Mats. 431, 172-184 (2012).
20. A. Dey, L. Jones, J. Nucl. Mats. 506, 63-70 (2018).
21. R. Garoby et. al. Phys. Scr. 93, 014001 (2018).
22. Los Alamos Scientific Laboratory. Group X-6. (1979)
23. M. Kawai, private communication.
24. T. T. Bohlen, F. Cerutti, M. P. W. Chin, et. al., Nucl. Data Sheets 120, 211-214 (2014)
25. A. Ferrari, P.R. Sala, A. Fasso, J. Ranft, CERN-2005-10 (2005) INFN/TC_05/11, SLAC-R-733
26. S. Saito, K. Suzuki, H. Obata, and Y. Dai, Nucl. Mat. Energy 34, 101338 (2023)
27. Y. Masuda, T. Kitagaki, K. Hatanaka, et. al., Phys. Rev. Lett. 89, 284801 (2002)
28. S. Ahmed, E. Altiere, T. Andalib, et al., Phys. Rev. C 99, 025503 (2019)
29. S. Ahmed, E. Altiere, T. Andalib, et al., Phys. Rev. Accel. Beams 22, 102401 (2019)
30. W. Schreyer, T. Kikawa, M.J. Losekamm, S. Paul, R. Picker, Nucl. Instr. Meth. A 858, 123-129 (2017)
31. W. Schreyer, C. A. Davis, S. Kawasaki, et. al. Nucl. Instr. Meth. A 959, 163525 (2020)

Ni target development for the TULIP project

Pascal Jardin[1]*, *Vincent* Bosquet[1], *Samuel* Damoy[1], *Georges* Frémont[1], *Franck* Perocheau[1] and *Marion* MacCormick[2]

[1]GANIL, Grand Accélérateur National d'Ions Lourds, Bvd H. Becquerel, BP55027 14076 Caen cedex5, France
[2] IJCLab, Institut Joliot Curie Laboratory, 15 Rue Georges Clemenceau, 91400 Orsay, France

Abstract. The TULIP project (https://anr.fr/Projet-ANR-18-CE31-0023) aims to produce radioactive ion beams of short-lived neutron-deficient isotopes using fusion-evaporation nuclear reactions in an Isotope Separator On Line (ISOL) system. A Ni target was chosen to produce Rb and Sn isotopes, and was bombarded, respectively, by Ne and Cr primary beams at energies close to the Coulomb barrier. Owing to the Target Ion Source System (TISS) configuration, the operational TISS temperature had to be close to 1300°C. Several tests were performed to determine a configuration able to cope with the constraints related to the TISS, target and beam. This article presents a brief description of the TULIP principles and objectives, and the development of the Ni target.

1 Introduction

The SPIRAL1 (Système de Production d'Ions Radioactifs Accélérés en Ligne, phase 1) isotope separator on line (ISOL) facility, installed at GANIL (Grand Accélérateur National d'Ions Lourds) has been producing radioactive ion beams (RIB) for 20 years. RIB, mainly from gaseous elements, have been produced using an electron cyclotron resonance ion source. Within the framework of an extended authorisation given to GANIL, a far greater variety of other RIB can be produced through combinations of current target materials and primary beams.

Since 2010, new TISS have been developed to enlarge the panel of RIB offered to GANIL users. The TULIP project started in 2015, addresses the production of short-lived neutron deficient isotopes. Its operational principle is quite different from mainstream TISS as it favours efficient atom-to-ion transformation (AIT) over in-target production rates. In the case of Rb^+ RIB production, which is the first case studied within the project, the target is a 4 µm thick Ni foil impinged with a Ne ion beam (4.5 MeV/A, ~1.5 pµA). At this energy in-target production is relatively low (~10^6 pps for ^{76}Rb for example). However, the compact TISS design allows to operate with greater AIT efficiency. Overall, the global RIB production should be high enough to satisfy user requirements, especially for short-lived isotopes.

The TULIP TISS is briefly presented before a more detailed description of the trials performed to obtain a satisfactory Ni target.

2 TULIP target constraints

TULIP essentially consists of a Ni target and a graphite catcher enclosed in a graphite cavity (Fig. 1). The initial aim was to maintain the cavity and the catcher at a temperature of 1573 K through ohmic heating, produced by high current circulating in the cavity walls and the target. Considering the case of a Ne beam on a Ni target, evaporated residuals, including Rb species, continue their post-production path through the Ni target up to the graphite catcher where they stop within a few µm. These radioactive atoms diffuse out of the catcher, effuse in the cavity, are surface ionized and finally accelerated to form an ion beam.

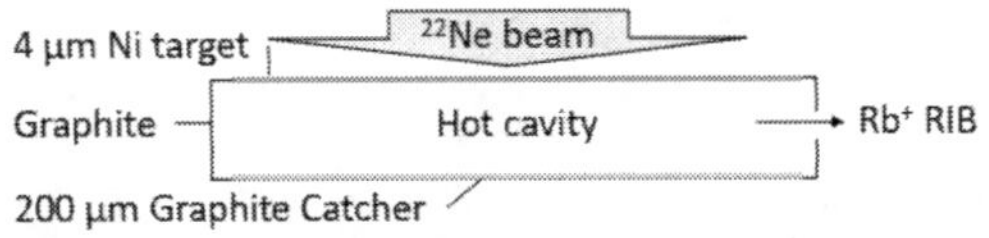

Fig. 1. Principle of the TULIP TISS.

To maximize in-target production while minimizing radiation and temperature damage induced on the target, the primary beam power is spread over a disc of 40 mm in diameter. To limit Rb atom sticking and effusion times, the cavity temperature must be maximized, while maintaining a target lifetime of at least 7 days.

3 First target heating tests.

The cavity was maintained at a temperature of 1620 K for 8 hours with a vacuum pressure between 10^{-6} and 10^{-7} mbar. The target's appearance progressively changed: The foil shrank and tore, and the surface changed from glossy to mat (Fig. 2). A SEM (Scanning Electron Microscope) observation (Fig. 3) revealed Ni crystals and fusion at the torn edges. At 1620 K pure Ni does not fuse but when exposed to carbon the fusion temperature drops to 1599 K [1] for 2% content by weight. Consequently, the cavity temperature must be limited to 1573 K.

* Corresponding author: jardin@ganil.fr

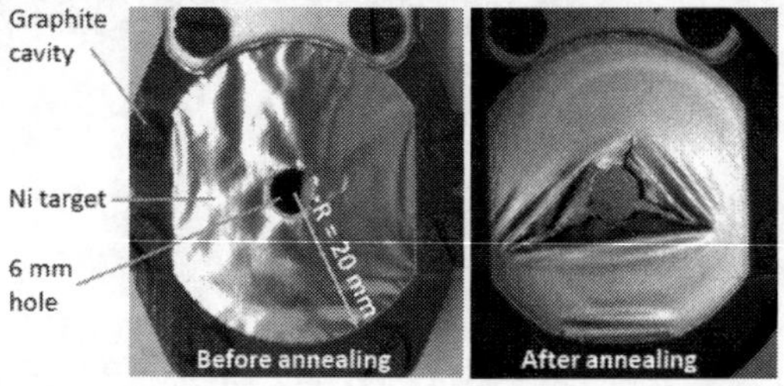

Fig. 2. Pictures of the Ni target mounted on the graphite cavity. The central hole is required to provide access to the catcher temperature measurement.

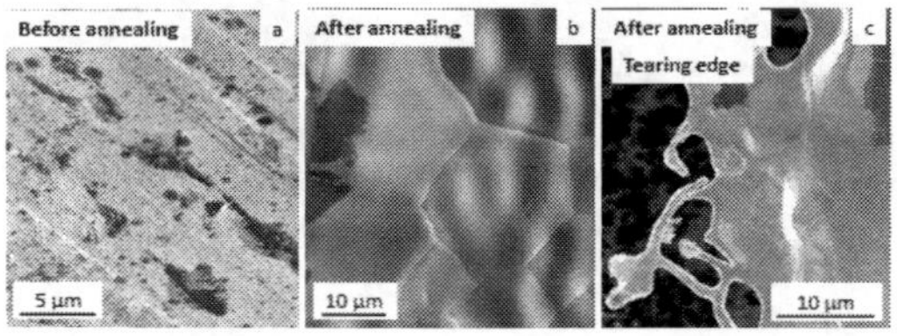

Fig. 3. SEM pictures of the Ni target before (a) and after (b, c) the heating period. Fig. 3a: Lamination grooves. Fig. 3b: Crystallisation. The black zone on the left corresponds to a thin carbon layer. Fig. 3c. Ni fusion on the torn edge. The dark background corresponds to the sample carbon support used in the SEM.

Comparable observations were made with different versions of the TULIP TISS during experiments aimed at not only characterizing the target material, but the overall TISS performances. The lifetime of the target was indeed longer at low temperatures. For temperatures of 1573 K and beyond, the maximum lifetime of the target was observed to be 8 h.

A more systematic study of the out-of-beam target behaviour, using progressive heat was performed. Each Ni foil was placed in a graphite oven and after each (i) heating period Δt_i (see Table 1), the total length of the foil was measured. A ~2×2 mm^2 sample was then taken from the foil and the average size of the grains measured.

Table 1. Heating test; duration, temperature, pressure versus sample number i, with observed relative shrinking and grain sizes. The calculated "annealing magnitude" (see text) is given in column 5.

Sample number	Outgasing time at 473 K	Heating temperature	Heating time	$\sum \lambda_i \cdot \Delta t_i$	Relative shrinking	Average grain size
i	hours	K	hours		%	µm
1	18.6	1079	2	3.4E-05	0	7.58
2	15	1122	2	1.7E-04	0	8.51
3	16.6	1243	2	3.8E-03	0	9.13
4	15.5	1319	2	2.7E-02	0.63	16.4
5	15.4	1468	2	2.3E-01	1.39	25.1
6	16	1558	2	1.3E+00	2.86	38.1
7	3	1566	15	1.8E+01	4.19	45.2

From the observed correlation between relative shrinking and grain size (Fig. 4), the shrinking process was expected to saturate for a heating time just beyond 15 h at a temperature close to 1573 K. This time was estimated by considering the material change was essentially due to the temperature induced movement of Ni atoms. To compare results, an "annealing magnitude" was defined as being the product of Δt_i and the kinetic rate λ_i, where λ_i depends on the temperature and on the activation energy for atomic displacement, according to the Arrhenius law [2].
According to the fit of relative shrinking versus $\sum \lambda_i \Delta t_i$, a maximal shrinkage should be obtained for $\sum \lambda_i \Delta t_i$ higher than 5.

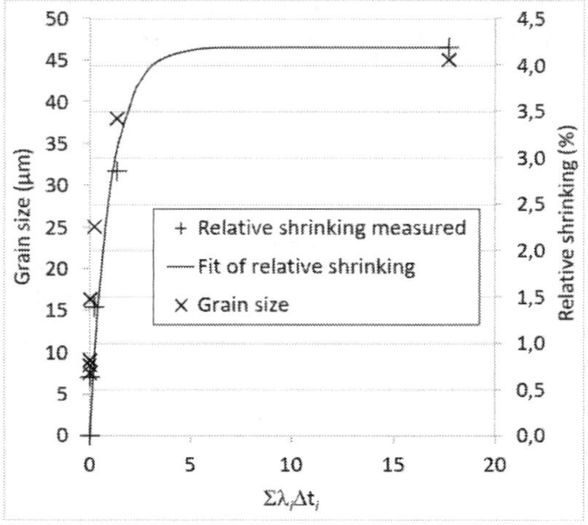

Fig. 4. Relative shrinking and grain size measured versus total annealing magnitude given by the sum $\sum \boldsymbol{\lambda_i \Delta t_i}$.

The aim of the measurement was to obtain a minimum relative shrinkage and mass evaporation, lower than 1%. Taking into account that λ_i depends exponentially on the temperature, the evolution obtained after 406 h at 1388 K, for which $\lambda_i \Delta t_i = 18.5$, should be comparable to the evolution obtained after 16 h at 1573 K. Another heat test was then performed to verify this assumption. Unfortunately, after only 31 h at 1388 K the target foil had shrunk by 10%, the surface became mat and the evaporation coefficient was higher than 4%. The assumption of shrinkage saturation after a given time thus seems incorrect and, in addition to the important evaporation rate, led to the abandon of the use of a pure Ni target even at a temperature of ~1373 K.

4 An alternative to a pure Ni

Adding another metal to Ni was considered, the intention being to reduce the mobility and evaporation rate of Ni atoms. Instead of using an alloy, a layer of a metal was considered. To avoid the production of isobaric contamination from nuclear reactions with the primary beam, the atomic mass of the added metal should be distant from that of Ni, and so excludes the transition metals of the fourth period in the Mendeleev classification. If the added metal is part of the sixth period, nuclear reactions with the primary beam can lead to alpha emitting nuclei, which then requires specific safety authorization at SPIRAL1. We therefore favored a metal from the fifth period.

To favor nuclear reactions of the Ne primary beam on mainly Ni nuclei, the relative abundance of the added metal must be minimal. The temperature induced phase changes in the alloy, due to beam wobbling, had to be avoided to limit mechanical stress.

The final choice is based on the two following observations. At a given temperature, the mobility (or self-diffusion) of metallic atoms in their own solid matrix decreases when their fusion temperature increases, as illustrated in Fig. 5.

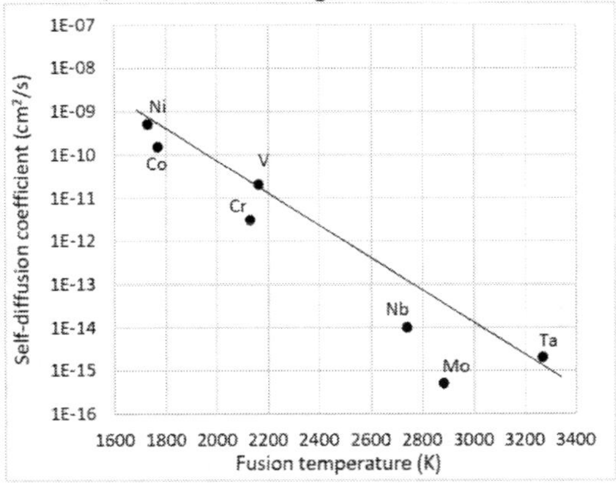

Fig. 5. Self-diffusion versus fusion temperature for metals of the fourth, fifth and sixth periods. Diffusion coefficients are extracted from [3].

Secondly, the fusion temperature and lattice parameter or density of the material are strongly related (Fig. 6).

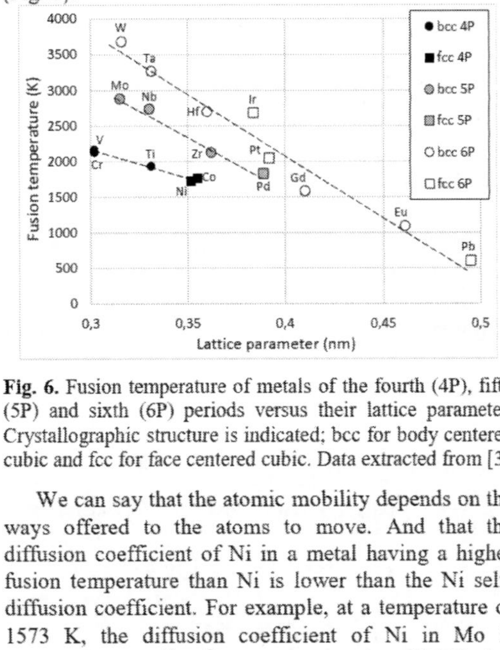

Fig. 6. Fusion temperature of metals of the fourth (4P), fifth (5P) and sixth (6P) periods versus their lattice parameter. Crystallographic structure is indicated; bcc for body centered cubic and fcc for face centered cubic. Data extracted from [3]

We can say that the atomic mobility depends on the ways offered to the atoms to move. And that the diffusion coefficient of Ni in a metal having a higher fusion temperature than Ni is lower than the Ni self-diffusion coefficient. For example, at a temperature of 1573 K, the diffusion coefficient of Ni in Mo is estimated to ~10^{-12} cm^2/s, compared to the self-diffusion coefficient of Ni of ~10^{-9} cm^2/s. Finally, within the fifth period of the Mendeleev classification, Mo seems to be the best candidate. Regarding the behavior with respect to temperature, the binary phase diagram given in ref. [4] indicates that for up to 20% in weight of Mo present in Ni-Mo alloy, the fusion temperature remains higher than 1693 K, and this fulfils the initial operational aim of 1573 K.

A 4 μm thick Ni foil was then covered with a layer of 0.09 μm of natural Mo on each face and was tested over 30 h at 1523 K. The relative shrinkage observed was lower than 1% and the relative mass evaporation was 0.8%, within the aim of 1% for a duration sufficient for a first irradiation test.

5 Mo-Ni-Mo sandwich target

Our observations of longer lived targets suggests that a thicker Mo covering is an appropriate choice. A 300 ± 6 nm thick layer was deposited on each face by KERDRY (https://kerdry.com) using the cathodic sputtering method [5]. The relative Mo weight was equal to 17%, corresponding to a fusion temperature of 1693 K.

The beam intensity to be supported was fixed to 1.5 pμA, corresponding to a primary beam power of 150 W, with ~30 W deposited in the target. As the target material inevitably transforms from multilayer to alloy due to heating, the emissivity of the target changes and is expected to be intermediate between the Ni (0.19) and the Mo (0.35) values [6]. Assuming the power is evacuated from the target only by radiation, the target temperature ranges from 1190 K to 1390 K. To reduce this temperature span, the target was placed 60 mm upstream from the cavity (Fig. 7), where the temperature induced by the cavity is negligible and estimated to be 640 K according to a simulation performed with the Ansys® software (https://www.ansys.com). To close the cavity, a 2 μm Diamond Like Carbon foil (https://www.micromatter.com) was used.

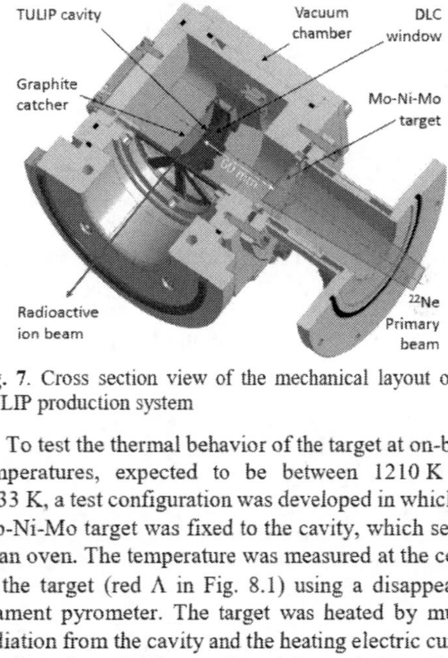

Fig. 7. Cross section view of the mechanical layout of the TULIP production system

To test the thermal behavior of the target at on-beam temperatures, expected to be between 1210 K and 1333 K, a test configuration was developed in which the Mo-Ni-Mo target was fixed to the cavity, which served as an oven. The temperature was measured at the center of the target (red A in Fig. 8.1) using a disappearing filament pyrometer. The target was heated by mutual radiation from the cavity and the heating electric current flowing through the cavity walls and through the thickness of the target, from right on Fig. 8 up to the exit situated on the left.

The current leaving the cavity is concentrated around the exit (Fig. 8.1), leading to higher temperatures (> 1300 K) and earlier target fusion. After 26 h (Fig. 8.2), the target is no longer connected to the cavity in the exit region, and the hottest points progressively move along the perimeter of the target. After 43 h, the target foil starts to ripple (Fig. 8.3 and 8.4) and stops rippling after ~65h (Fig. 8.5), when the current no longer has the possibility to flow through the target. Then the target shape stops changing (Fig. 8.6). This evolution and the aperture of the target explain the fluctuations of the measured temperatures and the variations of the heating current, which rose from 300 A to 350 A in order to maintain the temperature within the targeted range.

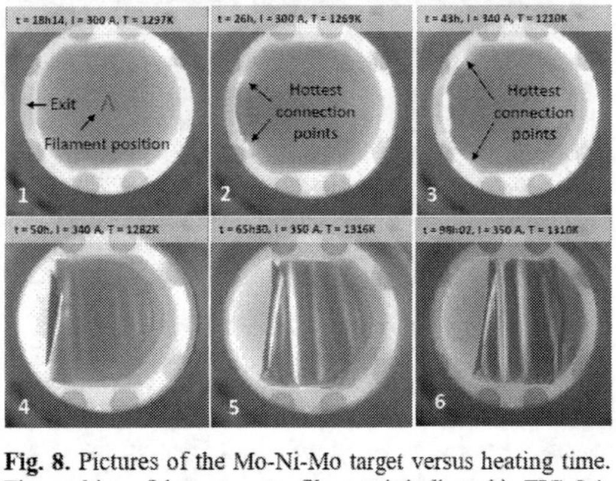

Fig. 8. Pictures of the Mo-Ni-Mo target versus heating time. The position of the pyrometer filament is indicated in FIG. 8.1. The relative brightness of the different pictures do not allow a comparison between temperatures as the pictures were taken with different camera settings.

After 98 h, the target foil was almost as glossy as it was before heating, suggesting weak crystallization. The flexibility was comparable to the initial one. The evaporated mass was equal to 10% and the relative shrinkage was lower than 1%. With respect to the Ni target requirements for TULIP, the present target seems to fulfil the requirements.

6 Target under ^{22}Ne primary beam

Conserving the mechanical layout as shown in Fig. 7 and the 0.3 μm Mo/4 μm Ni/0.3 μm Mo targeta ^{22}Ne ion beam impinged the target over 9 h at an energy of 4.5 MeV/A and a power up to 111 W (1.12 pμA or 7.0×10^{12} pps). The ion beam was wobbled on a 18 mm diameter with a 9 mm FWHM profile. The beam power density on the target was estimated to be 8 W/cm^2 at maximum.

After irradiation, tracks on the target revealed a clear misalignment of the beam on the target surface (Fig.9). In places where the beam misses the target, the temperature is estimated to be 640 K, and the target appearance did not change. On the positions corresponding to the FWHM beam diameter, the maximum beam power density is estimated to be 6 W/cm^2. The final appearance of the surface is slightly mat, revealing a change in the metal structure. On the rotating beam diameter, where the beam power density is maximum (~8 W/cm^2), part of the target has disappeared (the hole in Fig. 9).

At a beam power density of 8 W/cm^2, the estimated target temperature ranges from 1190 K to 1390 K. This temperature range cannot solely explain the formation of the hole as it is significantly lower than the fusion temperature of 1693 K deduced from the phase diagram given in ref [1]. Since the incident beam characteristics were not well controlled, excessive focusing and greater power density than desired cannot be excluded.

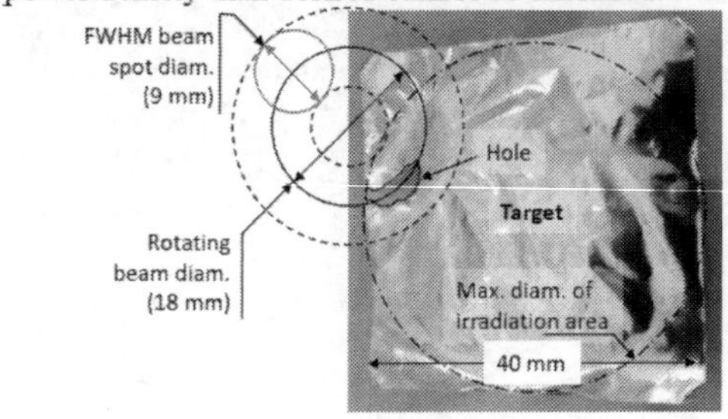

Fig. 9. Photo of the target after irradiation. The periphery of the hole is highlighted by a thin black line.

The temperature variations induced by the rotation of the beam are estimated to less than +/- 70 K, superimposed to an average temperature of ~1250 K, assuming a Mo weight percentage of Mo of 17%, an emissivity of 0.22, a heat capacity of 407 J.kg^{-1}.K^{-1}, an alloy mass density of 4.06×10^{-6} kg/cm^2, and an incident beam power density of 8 W/cm^2. With these conditions the operational working point should be in a region of the binary phase diagram where no material structure change occurs. Nevertheless, the operational parameter logbook shows some large beam power variations (from 10 W to 111 W), which could have created undesired mechanical stress provoking the formation of a hole.

From reference [7], the number of atoms sputtered from the Ni surface by $^{22}Ne^{+}$ ions should be less than 10, whereas a sputtering rate higher than 2000 is necessary to explain the formation of the hole. The effect of a sputtering created hole can thus be neglected.

Consequently, it is thought that excessive local power densities and beam intensity variations are the most probable contributors to the hole formation.

Acknowledgements
The TULIP project is supported by the French National Research Agency. The authors thank the GANIL groups, G. Renouf from CRISMAT laboratory and E. Gardès from CIMAP.

References

1. M.F. Singleton and P. Nash, ASM Handbook Alloy Phase Diagrams **3**, 533 (1992)
2. S. Arrhenius, Zeitschrift für Physikalische Chemie, **4**, 226 (1889)
3. G. Neumann and C. Tuijn, *Self-diffusion and impurity diffusion in pure metals*, Handbook of experimental data. Elsevier, Pergamon materials series **14** (2009)
4. H. Okamoto, Journal of Phase Equilibria **12**, 703 (1991)
5. S. Swann, Phys. Technol. **19**, 67 (1988)
6. V. Kuchi, PhD report, Uni. of CAEN, 159 (2018)
7. M. Toulemonde, in Proceedings of the Workshop on Materials in Extreme Environments, 13-14 May 2013, Michigan State University, USA (2013).

Manufacture of Tl targets by electrodeposition for the study of excitation functions of ^{203}Pb

Thomas Sounalet[1*], *Cyrille* Alliot[2], *Nadia* Audouin[2], *Nicolas* Bozovic[3], *Fabien* Brelet[2], *Remy* Dureau[3], *Arnaud* Guertin[1], *Ferid* Haddad[1,2], *Keerthana* Kamalakannan[1], *Etienne* Nigron[2], and *Maryne* Tarinas[2]

[1] Subatech, UMR 6457, IMT Atlantique, CNRS/IN2P3, Université de Nantes, 4 rue Alfred Kastler BP20722, 44307 Nantes Cedex 3, France
[2] GIP ARRONAX, 1 rue ARRONAX, CS10112, 44817 Saint-Herblain Cedex, France
[3] ORANO med, 2 route de Lavaugrasse 87250, Bessines-sur-Gartempe, France

Abstract. Natural Tl targets were manufactured by electrochemical deposition on a foil gold backing. The electrochemical parameters were defined after several experiments and reverse pulse potential was chosen to avoid the formation of filaments and dendrites. Once these parameters were established, enriched ^{205}Tl targets were manufactured on a gold foil backing to be used to measure the cross sections for ^{203}Pb production by deuteron induced reactions. The production yield was calculated from our excitation functions and was found to be 54 MBq/μAh in the energy interval 32.5 MeV – 30 MeV.

1 Introduction

^{203}Pb radiopharmaceuticals are good candidates for SPECT imaging [1] as ^{203}Pb emits a strong 279 keV gamma line and has a half-life of 51.9 h [2]. This interesting radionuclide forms the diagnostic part of the $^{203}Pb/^{212}Pb$ theranostic pair, whereas ^{212}Pb (10.65 h) [3, 4, 5] has been suggested for targeted alpha-therapy as it emits two β^- particles and one α particle during its decay chain. The production of ^{212}Pb is well known; it is eluted from a ^{228}Th generator, while ^{203}Pb can be produced by light charged particle (proton and deuteron) bombardment of natural thallium or enriched ^{203}Tl or ^{205}Tl. ^{201}Tl radiopharmaceuticals [6] [7] are widely used and ^{201}Tl is produced via a Pb precursor isotope from enriched ^{203}Tl with proton beam. The industrial fabrication of ^{203}Tl targets is achieved by electrochemical deposition using the method of bipolar chopped sawtooth form [6, 7]. This method requires a specific potentiostat/galvanostat, which is not always available in all industries or laboratories. A new method has been developed in this work based on electrodeposition. It has been used successfully to prepare thin targets to study the production of ^{203}Pb by deuteron irradiation of enriched ^{205}Tl as well as a thick target for large scale production. The stacked foil technique [10] was used to study the excitation functions of deuteron induced reactions on enriched ^{205}Tl targets. The present results are compared with Adam Rebeles et al. (2012) who used natural thallium [11] and Blue et al. (1978) who used enriched ^{205}Tl [12] in the energy range between 21 MeV and 34 MeV, and with the values extracted from the TENDL-2021 nuclear data library [8].

2 Materials and Methods

Thin deposits of natural Tl were obtained by electroplating on a high-quality gold foil (15 μm thick from Goodfellow, 99.90%). The plating solution was prepared with extreme caution due to the high toxicity of Tl. Beaker A, containing 1g Tl_2O_3, was filled with 0.5 mol/L EDTA, 0.4 mol/L NaOH, and 300 μL BRIJ-35 20%, then ultra-pure water was added to 33.1 mL. Beaker A was covered with Parafilm and the solution was stirred with a magnetic rod at 300 rpm. Beaker B was filled with 1.9 mL hydrazine hydrate 35%. The solution of beaker B was transferred through capillaries with a peristaltic pump at a flow rate of 0.5 mL/min to the solution of beaker A. The foam layer formed on top of the solution disappeared after 5 min.

The solution was adjusted to pH 8 by adding NaOH or H_2SO_4 as needed. This pH value was chosen because the Tl(I) and Tl(III) complexes with EDTA are stable at pH 8 [13], thus avoiding the precipitation of Tl(I) and Tl(III) hydroxide in the basic solution [14]. The electrodeposition occurred via the electrochemical reaction

$$Tl^+ + e^- = Tl$$

with a redox potential of -0.336 V/NHE [15]. Other states of thallium exist in the solution, such as Tl(III) that forms during electrodeposition. Tl(III) can be produced by the oxidation of Tl^+ on the auxiliary electrode to form thallium hydroxide or thallium oxide because of its redox potential of 1.252 V/NHE [15]. To

* Corresponding author: thomas.sounalet@subatech.in2p3.fr

avoid the formation of thallium oxide, hydrazine was used as a strong reducing anodic depolarizer.
All chemical products mentioned were purchased from Sigma Aldrich.

The deposition was carried out in a homemade Teflon cylinder cell containing 35 mL of the electrolyte solution. A 20 mm diameter hole was placed on the side of the cylinder for the working electrode resulting in a ^{205}Tl deposit of 3.14 cm^2. A sealing film and a stainless-steel plate were used to ensure a leak tight cell and good electrical connection. The working electrode was a 2.5 cm × 2.5 cm × 15μm gold foil. The auxiliary electrode was made from a 1 mm diameter platinum rod. A reference electrode Ag/AgCl/sat KCl was also used during the process. The distance between the working electrode and the auxiliary electrode was set at 1.0 ± 0.1 cm.

A PGZ 401 potentiostat/galvanostat was used to study the electrochemical behaviour at the interface between the solution and the substrate by voltammetry experiment, and to obtain different deposits by chronoamperometry according to voltammetic results following two different methods: one using direct potential and the other by reverse pulsing potential. The scanning potential was from 0 V/NHE to - 1.4 V/NHE for three cycles. The scanning speed was 0.05 V/s and the duration of chronoamperometry was set to 1h. For all experiments, the solution was stirred at 750 rpm using a homemade Teflon propeller rod.

The surface morphology of the deposits was analyzed by JEOL JSM 71OOF scanning electron microscope (SEM). Energy-dispersive X-ray spectroscopy (EDX) coupled to the SEM was also used to investigate the chemical composition of the surfaces. The stacked-foil technique [10] was used to perform production cross section measurements of deuteron induced nuclear reaction on enriched ^{205}Tl (Trace Science International). The target deposits were manufactured by electrodeposition as previously described. Experiments were conducted at GIP ARRONAX [16] in Saint-Herblain - France, in the energy range from 21 MeV to 34 MeV. The measured production cross sections were compared with the experimental data available in the literature (e.g., on the database EXFOR [2]) and from TENDL-2021 [17].

3 Results and Discussions

3.1 Voltammetry experiments

Figure 1 shows the voltammetry curves of Tl^+ in the solution at pH 8. The behaviour of the curves remained unchanged for three cycles. Two peaks occurred, one at the reduction of Tl^+ to Tl and one at the oxidation of Tl to Tl^+, at the potential of -1.0 V/NHE and -0.3 V/NHE, respectively. At -0.65 V/NHE, a small peak corresponds to the reduction of oxygen [18], which is inevitable. The curves drop at -0.8 V/NHE, which corresponds to the beginning of the reduction of Tl^+ to a Tl deposit. At -1.0 V/NHE, the value of current density is at a maximum, -6.8 mA/cm^2 and below -1.2 V/NHE, the curves are disturbed because of the reduction of water. At return to -0.8 V/NHE, the curves rise until -0.3 V/NHE, which corresponds to the dissolution of the Tl deposit. The maximum value of current density is 17.3 mA/cm^2 and it drops to 0 at -0.12 V/NHE.

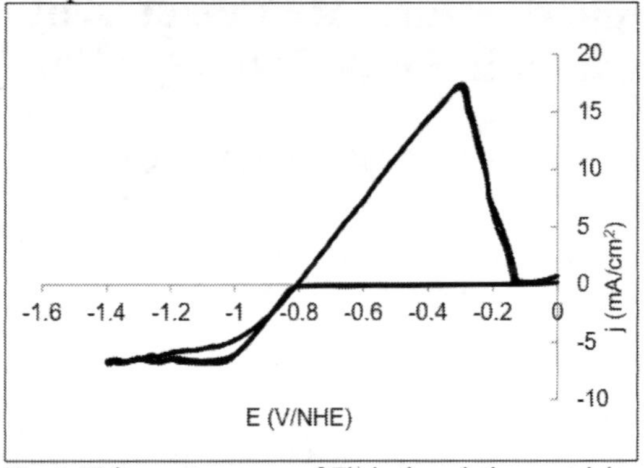

Fig. 1. Voltammetry curves of Tl^+ in the solution containing 0.05 M Tl_2O_3, 0.5 M EDTA, 1.9 mL hydrazine, and 300 μL BRIJ-35. The pH was adjusted to 8. Sweep from 0 V/NHE to – 1.4 V/NHE for 3 cycles with a scanning speed of 0.05 V/s.

3.2 Electrodeposition experiments: Direct potential

Five values of potential were applied: -0.6 V/NHE, -0.8 V/NHE, -1.0 V/NHE, -1.2 V/NHE, and -1.4 V/NHE. The deposits were characterized by SEM with the magnification x270, except for -0.8 V/NHE at x300 (Figure 2).

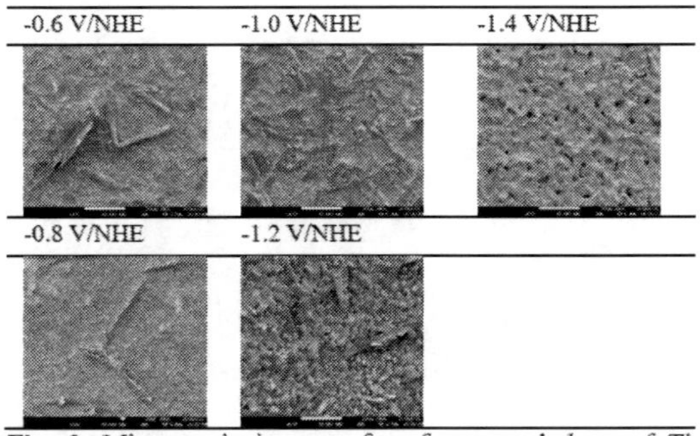

Fig. 2. Microscopic images of surface morphology of Tl deposits under different applied potentials: -0.6 V/NHE, -0.8 V/NHE, -1.0 V/NHE, -1.2 V/NHE, and 1.4 N/NHE.

For each condition in Figure 2, the formations of filaments are clearly visible on the surface of deposits for the potentials from -0.6 V/NHE to -1.2 V/NHE. At -1.4 V/NHE, vacuoles are present on the surface structure, which means that evolution of hydrogen gas bubbles occurred due to reduction of water. Furthermore, the surfaces of the deposits were porous, with the notable formation of dendrites. These defects make the Tl deposit fragile and prone to failure during irradiation. These defects have also been described by van den Bossche et al. [8].

https://doi.org/10.1051/epjconf/202328509001

3.3 Electrodeposition experiments: Reverse pulse potential

To avoid the formation of these defects, the reverse pulse current with a bipolar chopped sawtooth form mode was used to obtain a smooth and dense Tl deposit, as suggested for the electrodeposition of Tl for the production of ^{201}Pb from ^{203}Tl [9]. In our case, this involved replacing the current applied approach by the potential approach with a bipolar chopped sawtooth form. Van den Bossche et al. [8] had proposed working with a signal frequency of 100 Hz, a chopper frequency of 1000 Hz, and a chopper duty cycle of 60%. Our potentiostat/galvanostat did not allow application of high frequency, so 10 Hz was chosen. Furthermore, the pulse was accurate with this frequency using our potentiostat/galvanostat. After several experiments, we established a suitable potential to reduce Tl^{+}, while avoiding the formation of filaments and dendrites. The electrochemical parameters are summarized in Table 1, along with two pictures: morphology under SEM at x270 magnification, and Tl deposit on gold backing.

Table 1. Electrochemical parameters to obtain a smooth and dense Tl deposit

Reduction potential	Oxidation potential	Reduction time	Oxidation time
-1.10 V/NHE	-0.730 V/NHE	60 ms	40 ms

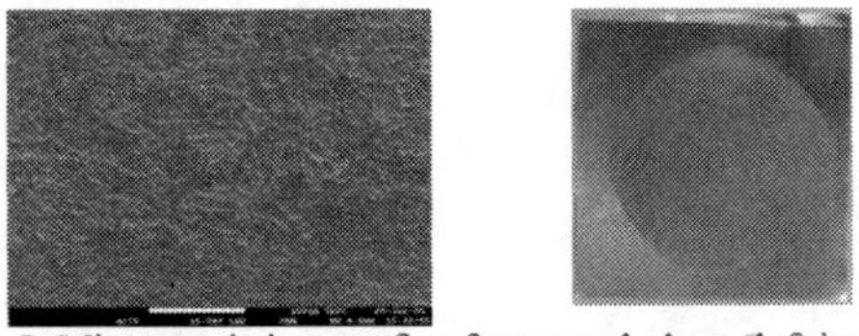

Fig. 3. Microscopic image of surface morphology (left image) and photograph of a Tl on gold backing (right image). The Tl deposit is obtained by applying a reverse pulse potential (-1.10 V/NHE for 60 ms and -0.73 V/NHE for 40 ms).

Figure 3 shows a smooth and dense Tl deposit free from filaments and dendrites. This deposit adheres strongly to the gold backing. The electrochemical parameters given in Table 1 were used to manufacture targets for measuring the excitation functions induced by deuterons on ^{205}Tl to produce ^{203}Pb.

3.4 Excitation functions of ^{205}Tl(d,4n)$^{203m1+m2+g}$Pb

Five irradiation runs were performed at the ARRONAX facility in the energy range 21–34 MeV with deuteron beams, using the stacked-foil method. Each irradiation involved two targets of ^{205}Tl deposits. The ^{205}Tl were manufactured by electrodeposition within 1h, to obtain a deposit thickness in the range 10 µm–15 µm.

Figure 4 reports the ^{205}Tl(d,4n)$^{203m1+m2+g}$Pb cumulative cross sections measured in this work in the energy range 21 MeV–34 MeV. The measured cross sections with the estimated total errors for this reaction at a certain energy are given in Table 2. Two existing studies in the literature are available: Blue et al. (1978) [12] and Adam Rebeles et al. (2012) [11]. The values of uncertainty from Blue et al. (1978) are not reported, and their values of cross section seem to be shifted in energy by a few MeV, leading to an underestimation below 28 MeV and an overestimation above 34 MeV. These values agree with the present work between 28 MeV–34 MeV. Despite the fact that in Adam Rebeles et al. (2012) natural thallium was used, the production of $^{203m1+m2+g}$Pb above the energy of 23 MeV is mostly due to the nuclear reaction induced on ^{205}Tl (natural abundance 70.48%). The values of cross section from Adam Rebeles et al. (2012) were rescaled to a 100% enriched ^{205}Tl material and are in good agreement with our data. The values available in TENDL-2021 [1] (dotted curve in Figure 4) agree well with our data for the energy below 30 MeV. At higher energy, these values are lower than our data and those reported by other authors.

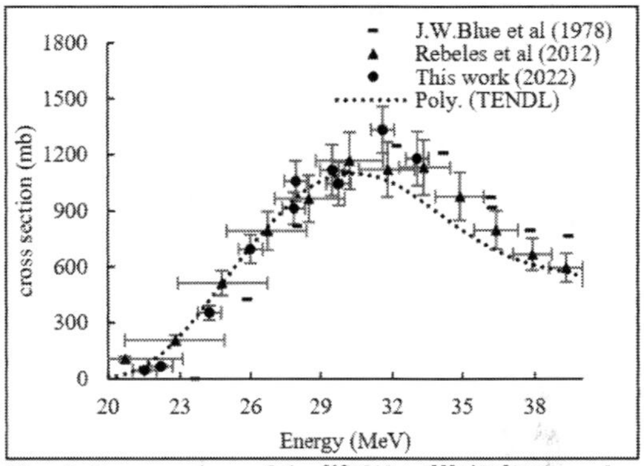

Fig. 4. Cross sections of the ^{205}Tl(d,4n)$^{203m1+m2+g}$Pb nuclear reaction

Table 2. ^{203}Pb measured cumulative cross sections of deuteron induced-reactions on enriched ^{205}Tl (enrichment of 99.75%).

E_d (MeV)	$^{203m1+m2+g}$Pb (mb)
21.5 ± 0.5	46.5 ± 5.2
22.2 ± 0.5	67.0 ± 6.9
24.3 ± 0.4	352.6 ± 39.1
26.0 ± 0.3	694.4 ± 77.2
27.8 ± 0.4	913.1 ± 86.2
27.9 ± 0.4	1058.7 ± 107.9
29.5 ± 0.3	1116.8 ± 136.8
29.8 ± 0.3	1044.5 ± 115.8
31.6 ± 0.3	1333.3 ± 124.3
33.1 ± 0.3	1180.0 ± 144.5

4 Conclusion

In this work, we determined optimized electrodeposition parameters to obtain a dense and smooth Tl deposit. Good quality deposits were obtained by the application of reverse pulse potential, which suppressed filament and dendrite formation. The same parameters were used to manufacture enriched ^{205}Tl targets for cross section measurements of $^{203m1+m2+g}$Pb through deuteron induced reaction in the energy range from 21 MeV to 33 MeV. Our data are in agreement with those reported by Adam Rebeles et al. (2012; within experimental errors), if scaling to 100% isotopic abundance is taken into account.

In general, the data available in the TENDL-2021 database show acceptable agreement with our experimental results.

Using our cross section data, we can estimate the ^{203}Pb production yield in the energy range from 32.5 MeV to 30 MeV to be 54 MBq/μA/h. This energy range corresponds to a thallium thickness of 42 μm.

Reference

1. E. Delpassand, I. Tworowska, R. Esfandiari, J. Torgue, J. D. Hurt, and R. Nunez, "Phase I dose-escalation study of AlphaMedix for targeted-alpha-emitter therapy of PRRT-naive neuroendocrine patients.," *JCO*, vol. 39, no. 15_suppl, pp. 4117–4117, May 2021, doi: 10.1200/JCO.2021.39.15_suppl.4117.
2. "National Nuclear Data Center." https://www.nndc.bnl.gov/ (accessed Jul. 03, 2018).
3. S. Hassfjell, "A ^{212}Pb generator based on a ^{228}Th source," *Applied Radiation and Isotopes*, vol. 55, no. 4, pp. 433–439, Oct. 2001, doi: 10.1016/S0969-8043(00)00372-9.
4. S. P. Hassfjell and P. Hoff, "A generator for production of ^{212}Pb and ^{212}Bi," *Applied Radiation and Isotopes*, vol. 45, no. 10, pp. 1021–1025, Oct. 1994, doi: 10.1016/0969-8043(94)90170-8.
5. G. L. Zucchini and A. M. Friedman, "Isotopic generator for ^{212}Pb and ^{212}Bi," *International Journal of Nuclear Medicine and Biology*, vol. 9, no. 1, pp. 83–84, Jan. 1982, doi: 10.1016/0047-0740(82)90082-1.
6. J. B. Elder, P. L. Tilkemeier, and J. A. Ewing, "Insights into Nuclear Cardiology in the United States from the first 3 years of the ImageGuide Registry," *J. Nucl. Cardiol.*, vol. 29, no. 1, pp. 166–176, Feb. 2022, doi: 10.1007/s12350-020-02112-5.
7. J.-Y. Huang *et al.*, "Improved diagnostic accuracy of thallium-201 myocardial perfusion single-photon emission computed tomography with CT attenuation correction," *J. Nucl. Cardiol.*, vol. 26, no. 5, pp. 1584–1595, Oct. 2019, doi: 10.1007/s12350-018-1230-y.
8. B. Van den Bossche, G. Floridor, J. Deconinck, P. Van Den Winkel, and A. Hubin, "Steady-state and pulsed current multi-ion simulations for a thallium electrodeposition process," *Journal of Electroanalytical Chemistry*, vol. 531, no. 1, pp. 61–70, Aug. 2002, doi: 10.1016/S0022-0728(02)01047-1.
9. "Cyclotron Produced Radionuclides: Principles and Practice," Feb. 28, 2019. https://www.iaea.org/publications/7849/cyclotron-produced-radionuclides-principles-and-practice (accessed Dec. 08, 2022).
10. G. Blessing, W. Bräutigam, H. G. Böge, N. Gad, B. Scholten, and S. M. Qaim, "Internal irradiation system for excitation function measurement via the stacked-foil technique," *Applied Radiation and Isotopes*, vol. 46, no. 9, pp. 955–960, Sep. 1995, doi: 10.1016/0969-8043(95)00179-H.
11. R. Adam Rebeles, P. Van den Winkel, A. Hermanne, F. Tárkányi, and S. Takács, "Experimental excitation functions of deuteron induced reactions on natural thallium up to 50MeV," *Nuclear Instruments and Methods in Physics Research Section B: Beam Interactions with Materials and Atoms*, vol. 288, pp. 94–101, Oct. 2012, doi: 10.1016/j.nimb.2012.05.044.
12. M. C. Lagunas-Solar, J. A. Jungerman, and D. W. Paulson, "Thallium-201 yields and excitation functions for the lead radioactivities produced by irradiation of thallium-205 with 34–60 MeV protons," *The International Journal of Applied Radiation and Isotopes*, vol. 31, no. 2, pp. 117–121, Feb. 1980, doi: 10.1016/0020-708X(80)90054-X.
13. J. Zembrzuska and B. Karbowska, "Identification of complexes involving thallium(I) and thallium(III) with EDTA and DTPA ligands by electrospray ionization mass spectrometry," *Rapid Communications in Mass Spectrometry*, vol. 31, no. 21, pp. 1785–1792, 2017, doi: 10.1002/rcm.7962.
14. T.-S. Lin and J. Nriagu, "Revised Hydrolysis Constants for Thallium(I) and Thallium(III) and the Environmental Implications," *Journal of the Air & Waste Management Association*, vol. 48, no. 2, pp. 151–156, Feb. 1998, doi: 10.1080/10473289.1998.10463658.
15. *CRC Handbook of Chemistry and Physics, 85th Edition*.
16. F. Haddad *et al.*, "ARRONAX, a high-energy and high-intensity cyclotron for nuclear medicine," *Eur J Nucl Med Mol Imaging*, vol. 35, no. 7, pp. 1377–1387, Jul. 2008, doi: 10.1007/s00259-008-0802-5.
17. "TENDL-2021 nuclear data library." https://tendl.web.psi.ch/tendl_2021/tendl2021.html (accessed Nov. 17, 2022).
18. Y. Z. Ussipbekova *et al.*, "Electrochemical Deposition and Dissolution of Thallium from Sulfate Solutions," *International Journal of Analytical Chemistry*, vol. 2015, p. e357514, May 2015, doi: 10.1155/2015/357514.

Discussions on liquid bismuth target use as an alternative for astatine-211 production

Théo Bigourdan[1]*, *Arnaud* Cadiou[2], *Arnaud Guertin*[2], and *Ferid* Haddad[1, 2]

[1]Arronax Nantes, 1, rue Aronnax CS 10112, F-44817 Saint-Herblain cedex
[2] Subatech - UMR 6457 - Laboratoire de physique subatomique et des technologies associées, 4 rue Alfred Kastler - La Chantrerie - BP 20722 44307 Nantes cedex 3 France

Abstract. Astatine-211 is an alpha emitter that has been identified as a good candidate for targeted alpha therapy. There is an increasing demand on this radionuclide. Very intense beam from linac being put in operation nowadays could be used to meet this demand. This document presents the design exploration of concepts of liquid bismuth targets dedicated to astatine-211 production. Three concepts are presented and analyzed: a capsule, a fluid loop and a windowless fluid loop. Structural and thermal sizing were performed using mechanical Finite Element models (ANSYS Workbench) and Computational Fluid Dynamic models (FLUENT). Production rates were assessed accordingly. Feasibility and expected performances are discussed in conclusion.

1 Introduction

While the use of radionuclides for diagnostic purposes (Single Photon Emission Computed Tomography, SPECT, and Positron Emission Tomography, PET) and therapeutic use of beta emitting radionuclides (iodine-131 or lutetium-177) are quite mature fields in nuclear medicine, the therapeutic use of alpha emitting radionuclides is still less developed. Targeted Alpha Therapy aims at using locally highly cytotoxic properties of alpha particles by carrying them to specific sites of cancerous cells thanks to vectors such as peptides or antibodies. This would improve the treatment efficiency, while reducing damages done to healthy tissues. Astatine-211 ($T_{1/2}$ = 7.2 h) is considered a promising candidate for such application thanks to its adequate half-life and its decay modes leading to the emission of one α particle by decay either directly or through its short lived polonium-211 daughter ($T_{1/2}$ = 0.515 s).

This work is part of the ANR-REPARE project that aims at improving astatine-211 availability for research and future medical application. This document focuses on target design for the latest accelerators [1, 2], like linear accelerators able to deliver alpha beam at very high intensities (order of magnitude of 1 mA).

2 Astatine 211 production

Astatine-211 is mainly produced using an alpha beam impinging a bismuth target, through the α+^{209}Bi reaction (energy threshold is 20.72 MeV). Among other parameters, beam energy must be carefully controlled to avoid production of astatine-210, which decays to the polonium-210, a long-lived ($T_{1/2}$ = 138 d) and highly cytotoxic bone seeker. Therefore, α energy for astatine-211 production must be in the 28.6 MeV to 20.72 MeV range [3].

3 Concepts

Solid target design for astatine-211 production must consider the low melting point of both bismuth (271°C) and astatine (approximately 302°C) and the high energy deposition rate of α particles [4]. To overcome these points, it was decided to start a concept study about a liquid bismuth target. Governing idea being to increase admissible temperature, beam intensity and production rate. To this end, three different concepts were evaluated: an encapsulated target, a flowing liquid bismuth target and finally a windowless fluid target.

3.1 Capsule design

3.1.1 Description

The capsule solution consists of a circular metallic receptacle, filled with bismuth (Figure 1). This kind of capsule is currently in service for other applications and material at the ARRONAX facility [1]. Before irradiation, the bismuth is in solid state and it will melt under the beam heat generation (in an uncontrolled manner). A cooling by pressurized water is considered on the back plate. An expansion volume will be included to absorb the bismuth's volume variations during melting and solidification.

* Corresponding author: theo.bigourdan@subatech.in2p3.fr

The following window structural materials were studied: stainless steel 316, havar and niobium. As a mean to improve mechanical resistance, curved windows were implemented in the model with various curvature radius.

3.1.2 Modelling

At this stage of the study, the model used for the beam power sizing was limited to the capsule itself using reasonable hypotheses regarding the cooling system and other interfaces. The window is the limiting factor regarding thermal and structural resistances. This window and its environment are thus modelled in different modules of the ANSYS Workbench software: Mechanical for structural resistance and Fluent for thermal and fluid considerations [5].

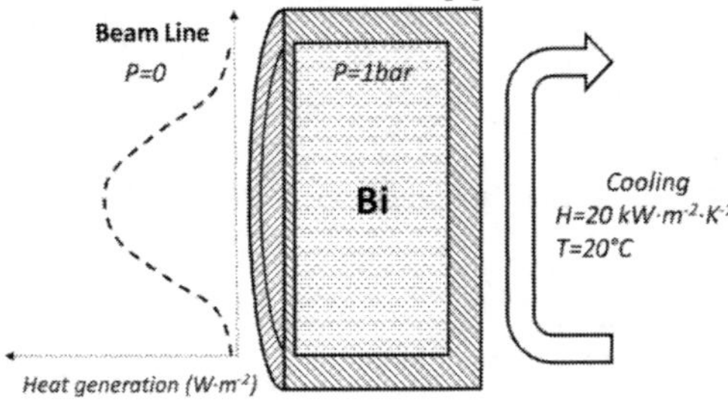

Fig. 1. Cross section of the capsule with thermal loads (H: Heat transfer coefficient, T: Temperature)

The window was modelled separately for the structural part. It is considered to be pinned on its external radius. Based on engineering judgement, pressure differential of 1 bar is considered between the beam line, which is under vacuum and inside the capsule.

As a conservative approximation, the beam heat generation is modelled by a surface heat flux on the external face of the window. The heat generation follows a Gaussian distribution as a function of the radius. The incident energy of the beam is fixed so that it enters the bismuth at 28.6 MeV to optimize astatine-211 production. The back plate cooling is modelled by a boundary condition with a heat transfer coefficient of 20000 $W \cdot m^{-2} \cdot K^{-1}$ and a temperature of 20°C, which is achievable with pressurized water.

The volume of bismuth being fully contained in the metallic structure, all the boundaries of the bismuth domain are considered to be fixed walls.

This configuration, summarized in Figure 1, allows us to perform optimizations for various materials and parameters.

3.1.3 Results

Table 1 shows the results of the computation for various configurations.

The havar having better yield strength than niobium, it can withstand the mechanical loads with a lower window thickness, which reduces heat generated for the same production rate. It gives the best result for capsule with a plane window. However, using havar would not have significant benefit with a curved window, since the melting temperature starts to be more limiting than the yield strength. The best-case scenario corresponds to niobium with a curved window and leads to an estimated production of 4.9 GBq in 1 hour. Since the astatine-211 available in the system does not increase linearly with time, the production in 1h was used as an indicator instead of the production rate ($Bq \cdot h^{-1}$).

3.2 Fluid loop concept

3.2.1 Description

This concept consists of a liquid lead-bismuth eutectic (LBE) circuit, fed by a pump and maintained in a liquid state by a thermal control system. A window (thinner pipe wall) is located where the beam impinges the flow. The flowing LBE contributes to dissipate the heat and act as a coolant for the window. A purging system and an in-line astatine-211 extraction system are considered as option if deemed necessary.

As the LBE only contains 55.5% of bismuth, the ^{211}At production is reduced by 44.5% compared to pure bismuth for the same beam characteristics. However, flowing liquid metal in metallic structures carries many complex issues that needs to be mastered for each situation and materials (corrosion, erosion, fragilisation…). The feedback from MEGAPIE [6] provides a good experience on flowing LBE in a stainless-steel structure. Moreover, the melting temperature of LBE (123°C) is lower than bismuth's, which reduces the stress on the structure and the risk of clogging. For the same reasons, it was decided based on engineering judgement to consider a minimal window thickness of 500 µm. According to the isotope inventory that was performed, the lead in the target does not create additional problematic isotopes compared to bismuth alone.

Tab. 1. Results of optimizations performed on window thickness and radius to maximize tolerable intensity

Capsule material	Window Radius	Optimal window thickness	Maximal Temperature	Initial Beam Energy	Maximal Beam Intensity	Production in 1 hour
-	mm	µm	K	MeV	µAe	GBq
Niobium (plane)	15	700	1269	76.7	63.9	1.4
Havar (plane)	20	200	1145	47.5	105.3	2.3
Niobium (curved)	20	100	1322	37.7	230.9	4.9
AISI316 (curved)	20	100	981	38.5	105.5	2.3

3.2.2 Modelling

At this stage of the study, only the surroundings of the window are modelled, as it is most likely to be the limiting factor for the production.

As per the capsule design, the window is modelled separately, in ANSYS Workbench. However, the pressure applied is the pump head pressure, estimated to 3 bars. This significantly increases the window stress.

Regarding thermal and fluid modelling, a tube portion is modelled, including a plane irradiation area. The structure and in particular, the window are not included. The beam heat is directly applied to the fluid beneath the window. Flow rate is set to get a mean speed of 1.2 $m{\cdot}s^{-1}$. [7]

3.2.3 Results

The maximal computed production rate is extremely low: 0.21 GBq in 1 hour. The thicker window, increased pressure and lower efficiency seem to jeopardize this solution by limiting production rate.

3.3 Windowless fluid loop

It has been established in previous concepts that the window is the limiting factor for beam intensity and astatine-211 production. In an attempt to bypass this technical limitation, the possibility of a windowless design was explored. The concept consists in irradiating directly an open flow channel of liquid LBE.

While removing the issue of the window, it rises some other questions that will need to be addressed.

3.3.1 Boiling and production rate

Boiling should be avoided as it may lead to excessive evaporation and projection of LBE in the beam line. In a first approach, this is the limiting factor for astatine-211 production.

Boiling temperature is reached when ambient pressure is equal to vapor pressure of the LBE at a given temperature. Ambient pressure is expected to be around 1 Pa as a preliminary assumption.

The LBE handbook [8] provides the following equation for the saturation vapor pressure:

$$P_{s(LBE)}[Pa] = 1.22 * 10^{10} * \exp\left(-\frac{22552}{T[K]}\right)$$

with T: Temperature (K) and Ps: Vapor pressure (Pa)

However, this formula may not be accurate since available experimental data are not consistent. Thus, to keep some margins we consider a LBE flow temperature of 885 K as a limit (which corresponds to a vapor pressure of 0.1 Pa).

LBE flow velocity in the beamline was assessed through an open flow channel analytic model and some Computational Fluid Dynamics correlations. It is assessed that the flow velocity could reach 3 $m{\cdot}s^{-1}$, in an optimistic configuration. By a subsequent thermal computation, it appears that an approximate production of 11.5 GBq in 1 hour can be achieved (14 kW of beam power). However, several issues need to be tackled.

3.3.2 LBE evaporation in the beamline

As the pressure around the free surface will be very low, the evaporation rate of the LBE must be considered. This sub-section aims at evaluating the corresponding flow rate and checking that sufficiently pressure can be maintained.

The LBE handbook [8] recommends the use of the Langmuir formulae to extrapolate the evaporation rate from the vapor pressure.

$$R_m = \sqrt{\frac{M}{2\pi RT}} p_{vs}$$

with M: Molar mass ($kg{\cdot}mol^{-1}$), R: Ideal gas constant ($J{\cdot}mol^{-1}{\cdot}K^{-1}$), T: Temperature (K), Pvs: Vapor pressure (Pa), Rm: Evaporation rate ($kg{\cdot}m^{-2}{\cdot}s^{-1}$)

As described in SLEEVE report [9], this formula is consistent with the rare experimental results available.

Gas flow was then evaluated based on the preliminary geometry described in Figure 2 and based on methods from [10]. Results summarized in the same Figure 2 show that the preservation of the pressure in the beam line is achievable.

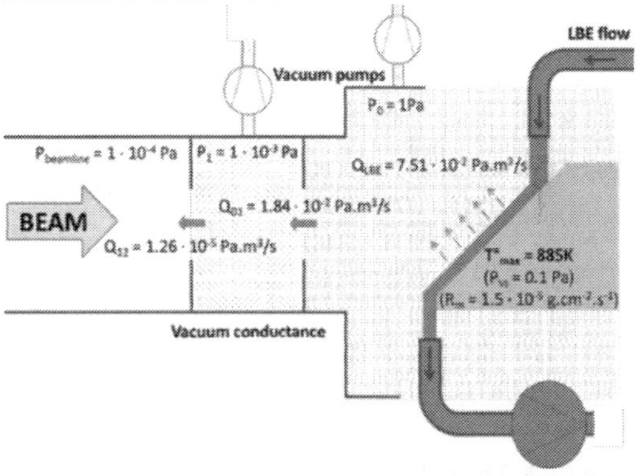

Fig. 2. Vacuum computation schematic and results (P: Pressure ; Q: Pressure flow ; Rm: Evaporation rate)

3.3.3 Astatine evaporation in the beamline

With this design, some extraction and losses of astatine-211 are foreseen. Our estimate relies on data provided by the LBE handbook [8] on iodine evaporation in LBE. As halogen, they may have similar behaviour. As there are no data available on astatine, it is the best assumption that we can use pending a dedicated astatine experiment.

We will consider that main losses occur by evaporation in the beamline. Losses and extraction rate will be assessed by assuming the evaporation rate in the vacuum.

$$R_m = \sqrt{\frac{M}{2\pi RT}} K_{H,I(LBE)} X_{At}$$

$$K_{H,I} = 10^{-\frac{10407}{T}+14.56}$$

with Rm: Evaporation rate ($g{\cdot}cm^{-2}{\cdot}s^{-1}$), M: Astatine atomic mass (kg), R: Ideal gas constant ($J{\cdot}mol^{-1}{\cdot}K^{-1}$), T: Temperature (K), K: Iodine Henry constant (Pa), X: Astatine molar fraction (-)

Evaporation is quite slow up to 700 K and almost instant at 1000 K. It may impose a criterion on maximal temperature in the irradiation area to avoid massive loss in the beamline.

Thus, based on these formulae and on simple time discretization, we performed a preliminary assessment on astatine-211 distribution between losses and extraction. We assessed that astatine-211 was uniformly distributed in the LBE volume and that the extraction occurs at the same time as the irradiation. The varying factor was the irradiation area temperature. Figure 3 shows the fraction of astatine-211 that can be retrieved (evaporated in the extractor) out of the remaining astatine-211 in the system (not decayed) as a function of said temperature.

Based on these results, we can assess the following:

- Evaporation rate increases greatly with temperature, with a major shift between 700 and 900 K.
- High temperatures in the irradiation area can lead to significant astatine-211 loss.

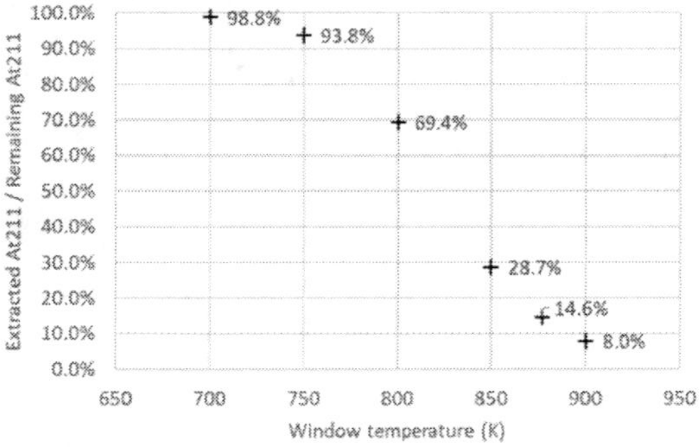

Fig. 3. Fraction of astatine-211 retrieved as a function of LBE temperature in the beamline

3.3.4 Astatine fixation to metallic structure

Halogens are known for binding to metallic compounds. As the LBE loop structure will be most likely made of stainless steel, there is a risk that an important part of the produced astatine-211 stays trapped there and cannot be retrieved. As per evaporation, there are no available data on the particular topic of migration of astatine from a LBE solution to a steel structure. However, we can find some data regarding iodine, the neighbor lighter halogen. Reference [6] provides with measurement of the distribution of iodine-129 in the LBE circuit of the MEGAPIE experiment (a liquid LBE spallation target).

The results shows that 93% of the iodine is located at the interface between LBE and steel. If the astatine behaves in a similar way, the concept of fluid loop may be jeopardized. While it is possible to extract astatine from steel, treating the whole LBE loop is unachievable. An experiment is under preparation to investigate this matter.

4 Discussion

Three metal liquid target concepts were described to produce astatine-211 and try to take advantage of the very high intensity α-beam that can be available in linac machine. Their efficiency and feasibility were assessed.

The interest of a very high-power beam dedicated installation should be questioned. Using reasonable technical means, it seems not feasible to use an intense beam (>1 mA) and a liquid target to produce large amounts of astatine. As shown in proposed concepts, it leads to important complexity, it is limited by physical constraints and it relies on knowledge on astatine behavior unavailable at this point. In addition, due to the half-life of astatine-211 an important proportion of production will be lost during transportation. Thus, a network of smaller regional production facilities (using solid targets or encapsulated liquid ones) should be investigated as it could be a more efficient way to provide the medical field with astatine-211.

Acknowledgment

A Grant from the French National Agency has supported this work for Research called Investissements d'Avenir, Labex IRON noANR-11-LABX-18-01, ISITE NExT no ANR-16-IDEX-007 and ANR-19-CE31-0013.

References

1. F. Haddad et al., *ARRONAX, a high-energy and high-intensity cyclotron for nuclear medicine*, Eur J Nucl Med Mol Imaging, **35**, 1377–1387 (2008)
2. S. Gales, *Spiral2 at GANIL: next generation of isol facility for intense secondary radioactive ion beams*, Nucl Phys A., **834,** 717c–723c (2010)
3. IAEA, *Charged-particle cross section database for medical radioisotope production*, Available: https://www-nds.iaea.org/medical/, Accessed in 2022 (2017)
4. G. Brunet et al., *Design of a high power solid target for ^{211}At*, EPJ Web Conf (to be published).
5. Ansys® Academic Research Mechanical, *Help System*, ANSYS, Inc., Release 2020 R2
6. B. Hammer-Rotzler et al., *Radiochemical determination of ^{129}I and ^{36}Cl in MEGAPIE, a proton irradiated lead-bismuth eutectic spallation target,* **103**, 745-758 (2015)
7. T. Kirchner et al., *LiSoR, a liquid metal loop for material investigation under irradiation*, Journal of Nuclear Materials, **318**, 70-83 (2003)
8. NEA, OECD, *Handbook on Lead-bismuth Eutectic Alloy and Lead Properties, Materials Compatibility, Thermal-hydraulics and Technologies* (2015)
9. K. Rosseel et al., *The Small LBE EVaporation Experiment*, SCK CEN (2009)
10. G. Rommel, *Production du vide*, Techniques de l'ingénieur (1977)

Novel solid target and irradiation methods for theranostic radioisotope production at the Bern medical cyclotron

Gaia Dellepiane[1*], *Pierluigi* Casolaro[1], *Alexander* Gottstein[1], *Isidre* Mateu[1], *Paola* Scampoli[1,2], and *Saverio* Braccini[1]

[1]Albert Einstein Center for Fundamental Physics (AEC), Laboratory for High Energy Physics (LHEP), University of Bern – Bern, Switzerland
[2]Department of Physics "Ettore Pancini", University of Napoli Federico II, Complesso Universitario di Monte S. Angelo – Napoli, Italy

Abstract. The production of medical radioisotopes for theranostics is essential for the development of personalized nuclear medicine. Among them, radiometals can be used to label proteins and peptides and their supply in quantity and quality suitable for clinical applications represents a scientific challenge. A research program is ongoing at the Bern medical cyclotron, where an IBA Cyclone 18/18 HC is in operation. The cyclotron provides 18 MeV proton beams up to 150 μA and is equipped with a Solid Target Station (STS) and a 6 m Beam Transport Line (BTL), ending in a separate bunker with independent access. A novel magnetic target coin was realized to bombard isotope-enriched materials in the form of compressed powders, together with a compact focalization system to enhance the irradiation procedure. For an optimized production yield with the required radionuclidic purity, novel methods were developed to precisely measure the extracted beam energy and the involved reaction cross sections. In particular, a target station was realized to measure nuclear cross sections using materials in the form of powder deposited on an aluminium disc by sedimentation, bombarded by a monitored flat beam.

1 Introduction

The availability of novel radionuclides is of paramount importance for the development of personalised nuclear medicine. For the theranostic approach, a pair of radionuclides with identical or very similar chemical properties is required to label the same biomolecule, combining diagnostic imaging and therapy. A diagnostic radionuclide (β^+- or γ- emitter) is used to detect possible diseases and to monitor the dose deposition imparted by means of the therapeutic radionuclide (β^--, Auger- or α- emitter), allowing to predict whether the patient will benefit from the treatment. For theranostics to be established in the clinical setting, a stable and sustainable supply of radionuclides in quantity and quality suitable for medical applications is required. The use of small medical cyclotrons and solid targets represents a promising solution. In this case, however, the precise knowledge of the beam characteristics during the irradiation is of paramount importance, and new instruments and methods are required. In this paper, we report about some of the developments and results obtained in the framework of a research program ongoing at the Bern University Hospital (Inselspital) cyclotron laboratory [1]. This research is aimed at optimising novel radionuclide production with solid targets. The facility hosts an IBA Cyclone 18/18 HC medical cyclotron (18 MeV proton beams, maximum current of 150 μA), equipped with 6 liquid targets for routine ^{18}F production, an IBA Nirta Solid Target Station (STS) and a 6 m long Beam Transport Line (BTL) that brings the beam in a second bunker with independent access to the experimental area. This solution allows multidisciplinary research activities [2-4] to be carried out in parallel with the industrial production of ^{18}F-labelled PET radiotracers.

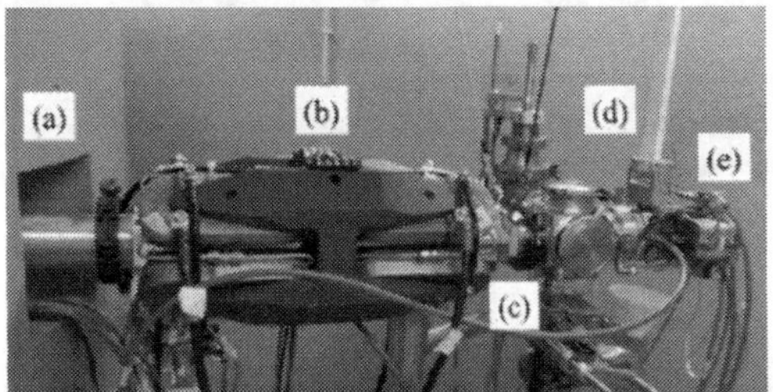

Fig. 1. The STS installed in the out-port of the cyclotron through the AFS: (a) cyclotron out-port; (b) Mini-PET beam line; (c) UniBEaM detector; (d) target receiving station of the hyperloop; (e) Solid Target Station.

2 Solid target developments

The STS is installed in one out-port of the cyclotron and allows to produce novel radioisotopes from materials in solid form. As shown in Fig. 1, the STS is connected to the cyclotron through the Automatic Focusing System (AFS), conceived by our group to enhance the

---

*Corresponding author: gaia.dellepiane@lhep.unibe.ch

irradiation performance [5]. The AFS is composed of the Mini-PET Beamline (MBL) [6], a two-dimensional beam profiler (UniBEaM [7]) and a software feedback system. The MBL is an innovative compact (50 cm long) pipe produced by the Canadian company D-Pace, embedding two quadrupoles and two steering magnets. The UniBEaM detector is a non-destructive two-dimensional beam profiler based on scintillating silica doped fibers passing through the beam. This detector was developed by our group and was successfully tested with a continuous 18 MeV proton beam in the current range from 1 pA to 20 μA [8]. A commercial version of the UniBEaM, shown in Fig. 2, is manufactured by D-Pace [9].

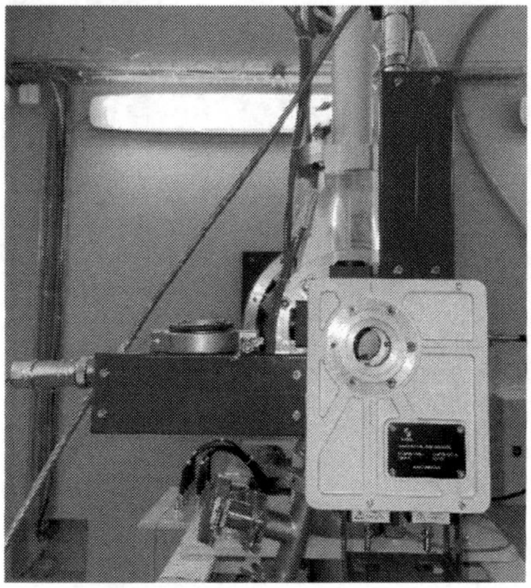

Fig. 2. Commercial version of the UniBEaM detector, produced by D-Pace.

The AFS steers and focuses the beam on target, controls its size and position and corrects its characteristics by means of the iterative feedback algorithm. This system allows to hit the target with the full beam extracted from the cyclotron, limiting overheating during the irradiation and unwanted residual activity, which would give rise to radiation protection issues.

During the irradiation, the target is contained in a special magnetic capsule, developed by our group to irradiate materials in the form of pressed powder pellets or solid foils. This capsule, called "coin" and shown in Fig. 3, is composed of two aluminium or niobium halves kept together by small permanent magnets. The thickness of the front part is used to adjust the energy of protons reaching the target material, in order to optimise the produced activity and the radionuclidic purity. The back part hosts the 6-mm-diameter target and an O-ring

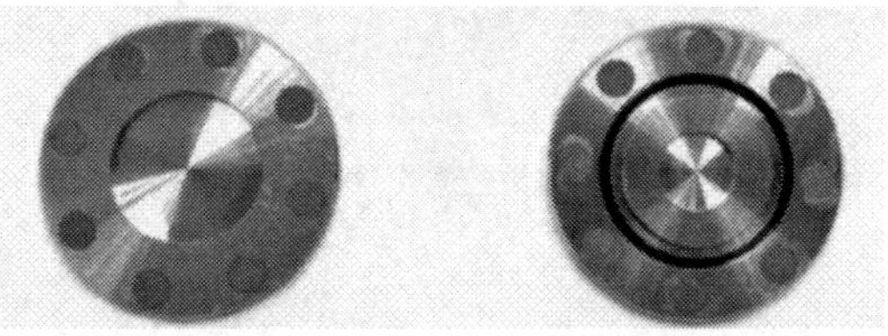

Fig. 3. The front and the back parts of the coin target (24 mm diameter, 2 mm thick).

to prevent the leakage of any molten material or gas produced during the irradiation.

The depth of the pocket containing the target can be varied from a few tens of μm (e.g. for irradiating materials in the form of thin foils) to about 1 mm.
To avoid overheating during the irradiation, the coin is water-cooled and helium-cooled, on the back and front sides, respectively. The body of the STS is connected to a B29885A Keysight electrometer to measure the current on-line during the irradiation. Radiochromic films [10] are used to measure the 2D beam profiles and allow to assess the effective current hitting the 6-mm pellet with an uncertainty within 10%.

Specific systems were implemented to load and unload the station without entering the cyclotron bunker, minimising the dose to personnel [11]. The target is inserted in the STS by means of a mechanical transfer system, called Hyperloop and shown in Fig.4-a, developed by our group. After the irradiation, it is transferred from the bunker area by means of the Solid Target Transfer System (STTS) by TEMA Sinergie (Fig. 4-b). The target can be delivered either to one hot cell in the nearby GMP radio-pharmacy, where the chemical processes are performed, or to a receiving station located in the BTL bunker (Fig. 5-a) for further analysis or for transport to external research laboratories. To assess the produced activity, the receiving station was equipped with a CZT detector system based on a ~1 cm^3 CdZnTe crystal (GBS Elektronik), shown in Fig. 5-b. The low counting efficiency makes this detector suitable for the high activities produced; moreover, its position can be modified by means of a programmable step monitor, up to a maximum of about 50 cm from the source. The CZT was experimentally calibrated with a High Purity Germanium (HPGe) detector, considering that the target always has the same size and lands with the same orientation. The system allows measuring the produced activity with an accuracy of a few percent [12]. Thanks to these developments, several radionuclides have been produced at the Bern medical cyclotron with solid targets, as reported in Table 1 [13]. In particular, the production of about 15 GBq of ^{44}Sc represents a very promising result in view of theranostic clinical applications [14].

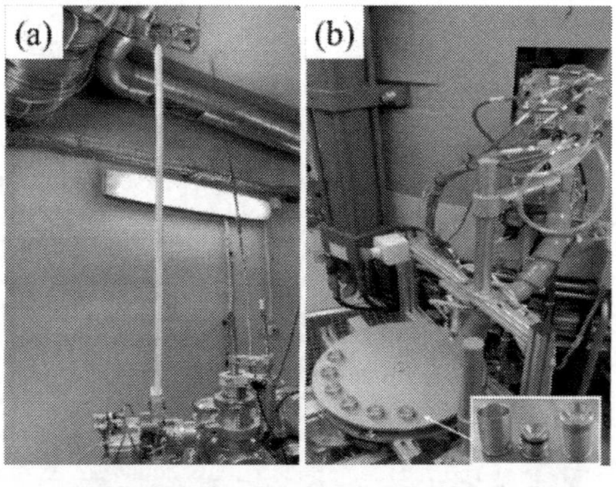

Fig. 4. (a) The Hyperloop connected to the STS; (b) the IBA Nirta STS and the STTS by TEMA Sinergie installed on the Bern medical cyclotron.

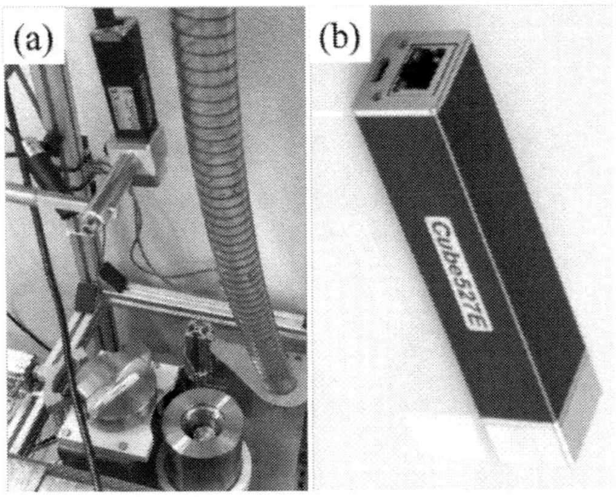

Fig. 5. (a) The receiving station located in the BTL bunker; (b) the CZT detector from GBS Elektronik.

Table 1. Main achievements in non-standard radioisotope production obtained with the STS at the Bern medical cyclotron.

Isotope	Reaction	Target	Mass [mg]	Y [GBq/μAh]
^{44}Sc	(p,n) (p,α)	enr. ^{44}CaO enr. $^{47}TiO_2$	30 65	0.6 0.05
^{47}Sc	(p,α)	enr. $^{50}TiO_2$	35	0.001
^{61}Cu	(p,α)	enr. ^{64}Zn	40	0.14
^{64}Cu	(p,n) (p,α)	enr. ^{64}Ni enr. ^{67}ZnO	63 59	0.13 0.02
^{67}Cu	(p,α)	enr. ^{70}ZnO	34	0.001
^{68}Ga	(p,n)	enr. ^{68}Zn	40	4.5
^{155}Tb	(p,n) (p,2n)	enr. $^{155}Gd_2O_3$ enr. $^{156}Gd_2O_3$	40 40	0.004 0.01
^{165}Er	(p,n)	nat. Ho	160	0.07
^{165}Tm	(p,2n)	enr. $^{166}Er_2O_3$	59	0.02
^{167}Tm	(p,n) (p,2n)	enr. $^{167}Er_2O_3$ enr. $^{168}Er_2O_3$	41 40	0.003 0.009

3 Beam energy measurements

The accurate knowledge of the proton beam energy is of paramount importance for an optimised production of novel radioisotopes for medical applications, where high radionuclidic purities are required.

In the out-port of the BTL, the extracted beam energy was accurately measured with different methods [15-17]. In particular, an apparatus based on the beam deflection by a dipole electromagnet allowed to study the beam energy as a function of several operational parameters of the cyclotron and provided a mean energy of (18.59 ± 0.14) MeV for a stripper angle of $97.1°$ [18].

Inside the cyclotron bunker, however, lack of space prevents the installation of experimental apparatus dedicated to the direct measurement of the beam energy. For this reason, the energy of the beam extracted to the STS out-port was assessed by means of a method based on the stacked-foils technique. The well-known monitor reaction $^{nat}Ti(p,x)^{48}V$ was chosen for this purpose, being characterised by a particular shape so that the produced activity strongly depends on the beam energy. Stacks of thin titanium layers, separated by 125-μm-thick niobium energy degraders were inserted in the coin pocket and irradiated. A specially designed coin with a hole in the front part was used to prevent the beam from degrading its energy. Each foil was then measured by gamma spectrometry with the HPGe detector, to precisely assess the produced activity. The measured activities were then plotted as a function of the Ti foil's number for several initial energies and an iterative least-squares minimisation method was applied to identify the initial energy that best reproduces the experimental data. The procedure was validated in the BTL and the beam energy measurement in the STS is currently in progress for different stripper angles. This methodology was developed by our group and can be easily applied to any medical cyclotron equipped with a solid target station, without the need of a beam transport line.

4 Cross-section measurements

To maximise the production yield of the desired radionuclide while minimising the production of any radioisotopic impurity, the precise knowledge of the cross sections of the involved nuclear reactions is of paramount importance.

A novel method was developed by our group to measure the cross sections using targets in the form of powder [19]. In most cases, the powder is deposited on an aluminium disc by sedimentation, which means that the target thickness cannot be precisely controlled. For this reason, the new procedure is based on the irradiation of the full target mass with flat proton beams, instead of the usual method based on the irradiation of a homogeneously thick target. The beam is flattened by the optical elements of the BTL and its position and shape are measured on-line with the UniBEaM detector. The beam current hitting the target is measured by means of a custom target station, shown in Fig. 6, composed of a 8 mm diameter collimator, an electron suppressor ring and a target holder. The collimator is grounded and provides a beam of controlled diameter, within which the beam flatness is achieved with an uncertainty within 5%.

The electron suppressor ring is connected to a negative bias voltage in order to repel secondary electrons produced during the irradiation, that would increase the measured beam current. Current measurements are performed on-line during the irradiation connecting the target holder to an ammeter (B29885A Keysight). The conductive parts of the target stations are kept together by insulator components. The beam is degraded by means of aluminium attenuator discs placed in front of the target and is determined using the SRIM Monte Carlo code [20]. The produced activity is assessed by gamma spectrometry with the HPGe detector.

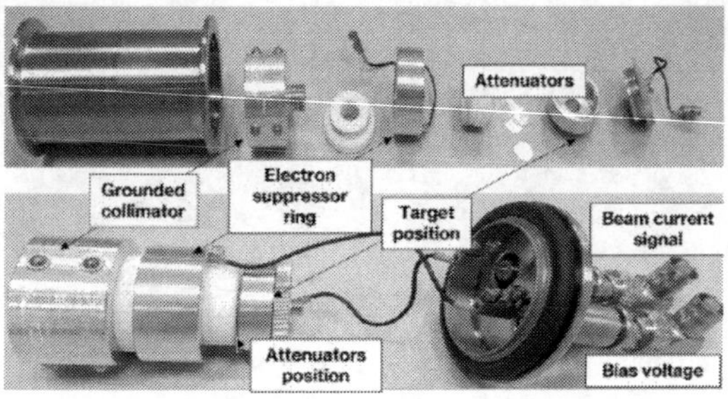

Fig. 6 Target station used for cross-section measurements.

Targets used for cross-section measurements consist of an aluminium 23 mm disc with a 4 mm diameter and 0.8 mm deep pocket in its centre. The target material is deposited by sedimentation method from a suspension of a few milligrams of material in ultra-pure water or ethanol, which is then evaporated. The target mass is assessed with an analytical balance (Mettler Toledo AX26 Delta Range). To prevent possible leakage of the material during the irradiation and the measurement procedure, the disc is covered by a 13 μm thick aluminium foil. With this procedure, target thicknesses of a few tens of μm can be achieved, and the beam energy can be considered constant within the uncertainties over the full irradiated mass. When the radionuclide production is the result of two or more nuclear processes, it is necessary to decouple their contribution to the production cross section. For this purpose, a method based on the inversion of a linear system of equation was developed [21]. It requires measuring the total cross section from as many materials with different isotopic compositions as there are reactions involved in the radionuclide production. With this method, the cross section of the nuclear reactions involved in the production of several radioisotopes (^{43}Sc [19], ^{44}Sc [19], ^{47}Sc [22, 23], ^{48}V [19], ^{61}Cu [24], ^{64}Cu [25], ^{67}Cu, ^{67}Ga [21], ^{68}Ga [21], ^{155}Tb [26], ^{165}Er [27], ^{165}Tm, ^{167}Tm and related impurities) were measured.

5 Conclusions and outlook

In the framework of the research program on the production of novel radioisotopes for nuclear medicine ongoing at the Bern medical cyclotron, new instruments and methods were developed. The irradiation procedure of solid targets was enhanced by developing a new target coin and a novel automatic focusing system. New procedures to measure the nuclear cross sections, the beam energy and the produced activity at EoB were also developed. The promising results obtained so far represent a valuable step toward the use of small medical cyclotrons for an efficient and reliable radioisotope supply in view of theranostic applications in nuclear medicine.

We acknowledge contributions from LHEP engineering and technical staff and from SWAN Isotopen AG team. This research was partially funded by the Swiss National Science Foundation (SNSF). Grants: 200021_175749 and CRSII5_180352.

References

1. S. Braccini, AIP Conf Proc **1525** (2013)
2. S. Braccini, *Compact medical cyclotrons and their use for radioisotope production and multi-disciplinary research,* in Proceedings of the 21st International Conference on Cyclotrons and their Applications (CYC2016), Zurich, Switzerland (2016)
3. S. Braccini et al., Sci Rep **12** (2022)
4. J. Anders et al., JINST **17** (2022)
5. P.D. Häffner et al., Appl Sci **11** (2021)
6. M. Dehnel et al., *An integrated self-supporting Mini-Beamline for PET cyclotrons,* in Proceedings of the 20th International Conference on Cyclotrons and their Applications (CYC2013), Vancouver, BC, Canada (2013)
7. S. Braccini et al., JINST **7** (2012)
8. M. Auger et al., JINST **11** (2016)
9. D.E. Potkins et al., Phys Procedia **90** (2017)
10. P. Casolaro, Appl Sci **11** (2021)
11. G. Dellepiane et al., EPJ Web Conf **261** (2022)
12. G. Dellepiane et al., J Phys Conf **2374** (2022)
13. G. Dellepiane et al., IL NUOVO CIMENTO C **44** (2021)
14. N.P. van der Meulen et al., Molecules **25** (2020)
15. K.P. Nesteruk et al., JINST **13** (2018)
16. K.P. Nesteruk et al., Instruments **3** (2019)
17. L. Campajola et al., Nucl Instrum Methods Phys Res B **440** (2019)
18. P.D. Häffner et al., Instruments **3** (2019)
19. T.S. Carzaniga et al., Appl Radiat Isot **129** (2017)
20. J.F. Ziegler and J.M. Manoyan, Nucl Instrum Methods B **35** (2013)
21. S. Braccini et al., Appl Radiat Isot **186** (2022)
22. T.S. Carzaniga and S. Braccini, Appl Radiat Isot **143** (2019)
23. G. Dellepiane et al., Appl Radiat Isot **189** (2022)
24. G. Dellepiane et al., Appl Radiat Isot **190** (2022)
25. G. Dellepiane et al., Appl Radiat Isot **191** (2023)
26. G. Dellepiane et al., Appl Radiat Isot **184** (2022)
27. N. Gracheva et al., Appl Radiat Isot **159** (2020)

Targetry for the in-beam activation of tracer particles for positron emission particle tracking

Thomas Leadbeater[1*], *Michael* van Heerden[1], *Andy* Buffler[1], *Ameerah* Camroodien[1], and *Gideon* Steyn[2]

[1]Metrological and Applied Sciences University Research Unit, Department of Physics, University of Cape Town, South Africa.
[2]iThemba Laboratory for Accelerator Based Sciences (iThemba LABS), Cape Town, South Africa

Abstract. Positron emission particle tracking measures the trajectory of a single radioactively labelled tracer particle by coincident detection of emitted annihilation photons. The technique enables the non-invasive study of dense opaque flows, with the tracer acting as a small neutrally buoyant flow-follower. The University of Cape Town has established a PEPT facility at iThemba LABS, utilising tracer particles produced through radiochemical methods, and measured using adapted positron tomographs. An activation approach producing the positron emitter ^{18}F inside glass target spheres of diameter between 5.0 and 10 mm using accelerated beams of alpha-particles has been explored. The reaction ^{16}O(α, x)^{18}F is used, exploiting the high concentration of natural oxygen and the correspondingly high cross-sections for ^{18}F formation. A standard target holder for the batch production of radionuclides at iThemba LABS was modified, reducing the entrance window thickness, allowing ingress of circulating cooling water, and adapted for a primary α-particle beam of 100 MeV energy delivered by the separated sector cyclotron (SSC) of iThemba LABS. Two-hour bombardment at nominal beam current 0.8 eμA produced activities up to ~ 110 MBq (3 mCi), with over 95% of the activity being ^{18}F.

1 Context

Positron emission particle tracking (PEPT) is a technique for measuring the trajectory of a single radioactively labelled tracer particle moving within an apparatus placed inside a PET scanner or positron camera [1]. For typical industrial applications, the tracer particle is usually a glass or ceramic bead, spherical in shape with a diameter between 5.0 and 10 mm, labelled with one of the medically significant positron emitters such as ^{18}F or ^{68}Ga. The coincident detection of 511 keV annihilation photons emitted by the tracer particle enables the non-invasive study of dense opaque flows and rapidly changing dynamic systems, including conditions of high turbulence [2]. The tracer particle acts as a small, neutrally buoyant, flow-follower in a viscous fluid, suitable to study fluid or granular behaviour in packed beds, tumbling mills and/or other flow systems [3-5]. The PEPT technique is particularly useful in studying well-mixed systems where tracking a single particle over an extended duration provides essentially the same information as simultaneously observing the instantaneous motion of all the particles in the fluid – formally referred to as ergodic systems [6]. Analysis of the single-particle trajectory over long timescales (minutes to hours) can provide time-averaged maps of velocity fields, the shape of dense flow regions, identification of regions of low-velocity flow, observation of gravity-dominated flows, and etc. [6].

The Department of Physics at the University of Cape Town (UCT) established a PEPT facility at iThemba LABS in 2009, based around a Siemens ECAT HR++ PET scanner modified for PEPT acquisition and analysis [7]. Alternative positron camera systems have since been added, including large area planar detectors (an ADAC Vertex dual-headed gamma-camera) and modular positron camera systems with customised detection geometries designed for specific applications, based on the block detector design [8]. Positron cameras formed from modified positron emission tomography (PET) scanners are used to detect and reconstruct the annihilation photon flight path, by detecting photon pairs in time coincidence. Genuinely reconstructed paths (which have not been produced by random coincidence, or corrupted by attenuation and scattering) converge upon the instantaneous position of the tracer particle. An iterative least squares minimisation approach is used to determine the 3D position where the photon paths converge, and hence the most likely instantaneous position of the tracer particle which can typically be determined at sub-millimetre precision at kilohertz rates. Tracking of particles moving with velocities up to 10 m/s is therefore achievable [7].

The tracer activity required for optimum tracking depends upon the amount of attenuating and scattering materials, and on the flow conditions, which vary considerably between applications. Typical tracer particles are required to range in activity from around 10 to 150 MBq (250 μCi to 3 mCi), be predominantly positron emitters, with ideally a short positron range. The principal tracer isotope used at PEPT Cape Town has been ^{68}Ga (half-life 68 minutes, 88.91% β^+)

* Corresponding author: Tom.Leadbeater@uct.ac.za

extracted chemically from the standard medical $^{68}Ge/^{68}Ga$ radioisotope generators produced by iThemba LABS. The ^{68}Ga is regularly eluted from the SnO_2 generator column, purified and concentrated, and used to radiochemically label suitable particles to be used as tracers within the study application. If a particle from the bulk material under study cannot be labelled chemically, an analogue particle is produced by attaching sufficient ^{68}Ga activity onto a resin substrate which is further modified to match the physical properties of the material in question. The representation of the tracer particle to the bulk material under study is critical, where systematic differences in physical properties can alter the dynamic behaviour and hence interpretation of PEPT results. The physical properties that are carefully controlled depend on the application and associated research questions, and include the physical (size, shape, surface roughness, density, etc.) and chemical (surface potential, wettability, hydrophobicity / hydrophilicity, etc.) attributes. When studying liquids, the radioactivity is sealed onto the substrate, and the overall particle density modified to match that of the fluid and achieve neutral buoyancy [9].

In the context of the accelerated beams and infrastructure available at iThemba LABS, an experimental campaign was initiated to activate silicon dioxide (SiO_2) based glass beads using high energy alpha particle beams. The method utilises the set of $^{16}O(\alpha, x)^{18}F$ reactions, exploiting the high concentration of natural oxygen in the glass as well as the relatively high cross-sections for the formation of the positron emitter ^{18}F. The competing (α, x) reactions on natural silicon, oxygen, and other contaminant species within the target matrix, have been shown to yield no long-lived radioisotopes instead producing short-lived (maximum half-life 2.5 minutes) or stable activation products. The activation is expected to yield ^{18}F of relatively high radioisotopic purity. Advantageously, ^{18}F has a longer half-life (110 minutes), higher positron branching ratio (96.73%) without any associated gamma emissions, and shorter positron range compared to ^{68}Ga implying longer experimental timescales at higher spatial resolution.

2 The $SiO_2(\alpha, x)$ reactions

The alpha particle induced reactions occurring with both naturally occurring silicon (92.22% ^{28}Si, 4.69% ^{29}Si, and 3.09% ^{30}Si) and naturally occurring oxygen (99.76% ^{16}O, 0.04% ^{17}O, and 0.21% ^{18}O) result in elastic and inelastic scattering and predominantly short-lived or stable activation products. The only significant long lived activation product is ^{18}F produced by competing alpha particle induced reactions on ^{16}O within the target [9]. The reactions, in order of significance given by relative total, cross-section are direct production by $^{16}O(\alpha, np)^{18}F$, $^{16}O(\alpha, d)^{18}F$, and indirect production by $^{16}O(\alpha, 2n)^{18}Ne \rightarrow {}^{18}F$ which proceeds by β^+ decay of the ^{18}Ne. The cumulative excitation function of the $^{16}O(\alpha, x)^{18}F$ reaction peaks at about 35 MeV, with a threshold at 18 MeV beam energy. Figure 1 shows the microscopic cross-sections for the three significant ^{18}F producing reactions, their sum, and the total cross-section for alpha particles of energy in the range up to 100 MeV, taken from TALYS calculations given in the TENDL database [10].

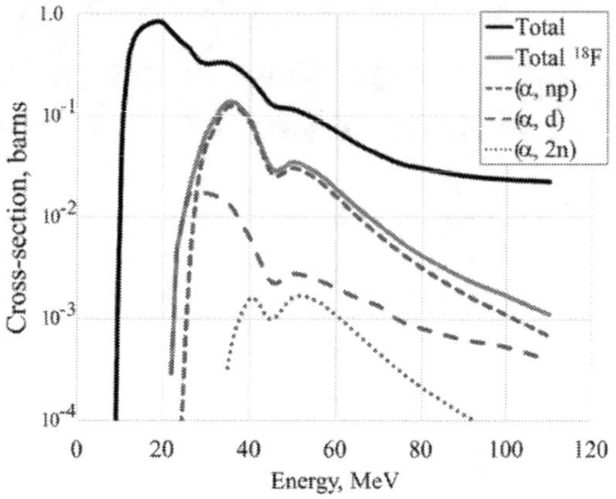

Fig. 1. Cross-sections for the $^{16}O(\alpha, x)$ reaction channels.

3 Target systems

Glass spheres (molar composition roughly 61% O, 23% Si, 10% Na, 3% Ca, 2% Al, 2% B, 1% Mg) of diameters 5.0 to 10 mm were used as targets, detailed composition can be found in references [8] and [9]. A target holder and capsule system were designed to deliver α-particles accelerated by the iThemba LABS separated sector cyclotron (SSC) to the targets and to maintain integrity by providing circulating cooling water. A standard target holder used for the batch production of radionuclides at iThemba LABS using the horizontal-beam target station 1 ("Elephant") [11] was modified for this purpose. The entrance window thickness of the target holder and the wall thickness of the target capsule were reduced. The target capsule was placed in the "high energy slot" inside the holder. Normally, in the case of radionuclide production, the capsule serves to protect the target material from direct contact with the cooling water. When activating glass beads, direct contact of the target material with the cooling water is essential to reduce the effects of beam heating. Deformation by swelling and/or melting potentially affect the representation and non-invasiveness of the tracer particle when performing PEPT. The target capsule was therefore modified by drilling a matrix of small holes through the front and side walls through which cooling water can freely flow. A further advantage of this approach is that the coolant (10 bar, 30 l/min flow rate) serves to agitate the target spheres in situ, where rotation allows beam entrance from different directions relative to the sphere centre, effectively isotropically irradiating the targets. Depending on the beam energy and target attenuation this approach enables the entire surface of the target sphere to be irradiated, rather than just the forward-facing surface as when using a fixed target.

Due to dead layers upstream of the target (including entrance windows, cooling layers, etc.) a significantly higher beam energy is required to reach the glass surfaces with the desired energy. The target holder and capsules were designed for a primary α-particle beam energy of 100 MeV. The beam passes two 25 μm thick Havar (ICRU-470 / UNS R30004) foils separated by a 10 mm helium cooling layer (maintained at 1.2 bar), a further 4.0 mm air gap, and aluminium entrance window of 0.5 mm to enter the target station, as shown in figure 2. For a 100 MeV α-particle beam extracted from the SSC, the beam energy impinging on the target surface is around 80 – 70 MeV determined by Monte Carlo calculation. The typical linear range of these alpha particles within the target material is around 3 mm, with energy above the reaction threshold within the first 2 mm of penetration. For small diameter targets (< 4 mm), the radioisotope activation is expected to be volumetric, with larger targets expected to activate to a surface layer of uniform depth, provided they are rotated under the coolant flow.

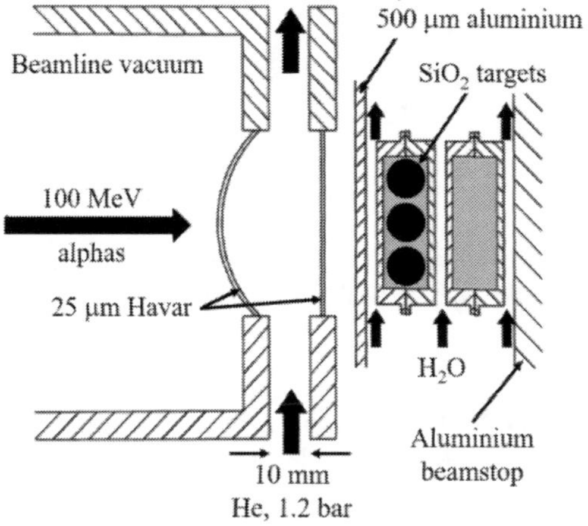

Fig. 2. Target stack, including dead layers.

4 Method

Multiple target spheres are placed within a single target capsule as shown in figure 3 and activated with the beam defocussed and swept in a circular pattern to increase the beam-strike area. Practically, this approach leads to a redundancy in activated targets and reduces the heating load per sphere, but not all receive the same amount of accumulated charge. A distribution of radioactivity (and heating damage) is then expected, from which a suitable candidate can be selected for the PEPT measurement.

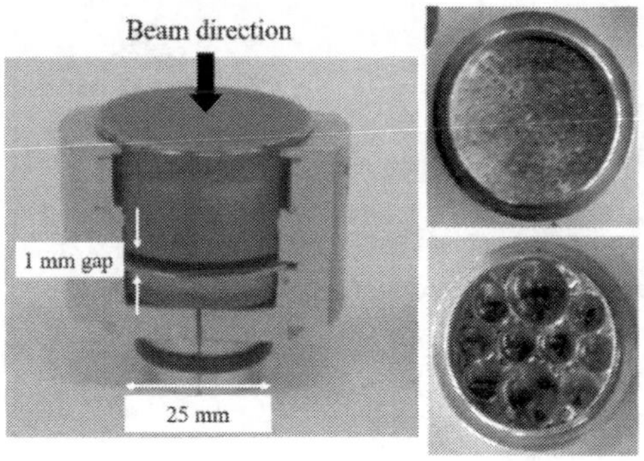

Fig. 3. Target capsule, perforated containment, and open capsule holding SiO_2 target spheres.

Several bombardments were performed with 100 MeV α-particles delivered by the separated sector cyclotron (SSC) of iThemba LABS. Helium gas was ionised in the ECR ion source and accelerated to 8 MeV for injection into the SSC, where the beam is further accelerated to 100 MeV. The extracted beam was directed to the target station through a set of dedicated beamlines and a $k = 100$ switching magnet, which limits the maximum beam energy delivered to the target in practice. Approximately two-hours bombardment at an average beam current of 0.8 eμA was sufficient to deliver 2.0 μAh of integrated charge, at which point the bombardment was terminated. A short cooling period (20 minutes – 1 hour) following end-of-beam allowed any short-lived radioisotopes to decay, and the target holder and capsule were recovered from the target station using robotic manipulation. The target spheres were unloaded from the capsule in a hot cell and transferred to the PEPT laboratory for activation product characterisation and activity measurements.

5 Results

Characterisation followed multiple simultaneous measurements using a wide range of detector techniques and extended over multiple half-lives of the produced species [9]. A calibrated high purity germanium (HPGe) spectrometer (energy resolution 1.2% at 511 keV) was used to determine the gamma radiation spectrum emitted by the targets. The most intense peak was observed at 510.5 ± 2.6 keV with associated sum peaks and Compton continua, corresponding to the positron annihilations expected from the decay of proton rich species. Based on the HPGe measurements, and assuming the yield of positron activity due to predominantly ^{18}F, a Capintec CRC25-R 4π ionization chamber calibrated to ^{18}F was used to measure the absolute activity of each sample over many hours, allowing the decay curve to be reconstructed and projected back to the time at end-of-beam. Ionisation chamber results from four independent bombardments, and 24 samples total, are shown in figures 4 and 5.

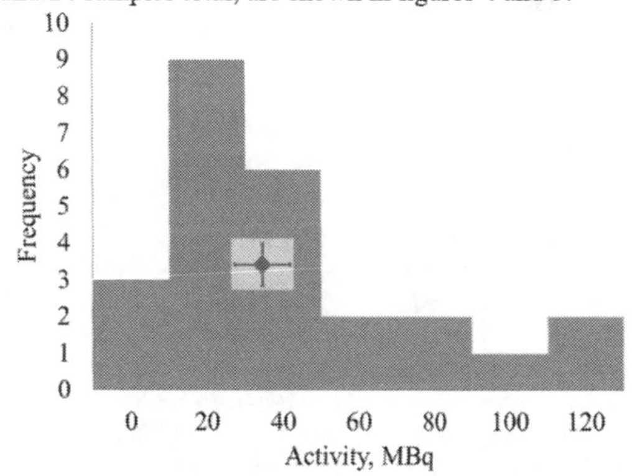

Fig. 4. Measured frequency distribution histogram of target activities, the point shows the mean for all samples.

Three target spheres had no appreciable activity above background, implying the beam did not impinge upon them with sufficient energy for activation. Three of the samples had measured activities above (75 MBq) (2.0 mCi), with the maximum activity around 110 MBq (3 mCi). The mean activity was 30 MBq (850 μCi) measured for all 24 samples. Time series gamma-ray spectroscopy using a calibrated sodium iodide (NaI) spectrometer enabled half-life measurements for each of the primary photopeaks, adding a high degree of confidence in their likely origin. Singles counting windowed around the 511 keV photopeak (350 – 850 keV) using 34 independent detector modules of the HR++ positron camera, and simultaneous prompt and delayed time coincidence measurements (12 ns coincidence gate) were used to calculate the absolute activity for each of the samples, confirming the ionisation chamber measurements [9]. The consolidated activity data for all samples are shown in figure 5 as a function of the squared radius of the target spheres, where the linear trend shows that expected from activation of the surface layer implying that the targets did undergo some rotation within the beam.

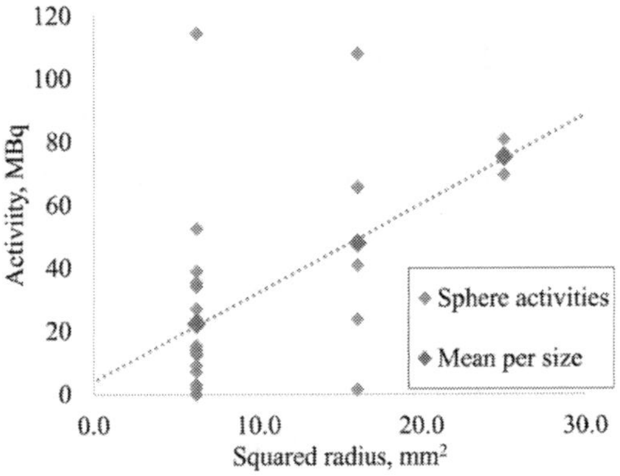

Fig. 5. Mean activities per target size.

Low intensity photopeaks measured with the HPGe spectrometer at energies 372.3 ± 1.1 keV and 1367.1 ± 1.7 keV were observed. By combining these measurements with the time series decay data per peak, these were attributed to activation of target composition admixtures. The positron emitter ^{43}Sc (511 keV (annihilation), 176.2% and 372.9 keV, 22.5%) was produced by alpha particle interactions with the 3% calcium admixture. The neutron rich ^{24}Na (14.98 hour half-life, principle gamma transitions at 1368.6 keV 99.99% and 2754.0 keV 99.85%) was produced by thermal and fast neutron interactions with Na, Mg, and Al, within the glass targets, present in ~1% amounts. For PEPT, the additional positron yield from ^{43}Sc is beneficial, extending the possible experimental timescale. ^{24}Na is not a positron emitter; however, annihilation photons are observed from this isotope due to the high energy photon emissions undergoing pair production. These are a likely source of background in PEPT, although the intensity is insignificant compared to the yield from positron emitting species. If these activation channels were to be used to produce ^{18}F for alternative applications (e.g. medicine), the observed contaminants can be controlled by modifying the purity of the target. Table 1 combines all measurement analyses and presents the yield of observed activation products projected back to the EOB.

Table 1. Relative activities of identified products at EOB.

Isotope	Decay mode	Half-life, hours	Relative activity, %
^{18}F	β^+ (96.73%)	1.8	~ 95%
^{43}Sc	β^+ (88.1%)	3.9	< 5%
^{24}Na	β^- (99.99%)	14.9	< 5%

6 Concluding remarks

The ^{16}O(α, x)^{18}F reactions using 100 MeV alpha particle beams at nominal intensity 0.8 eμA have been investigated in producing ^{18}F within 5.0 – 10 mm diameter glass targets of naturally occurring isotopic composition. The observed purity of ^{18}F was consistently 95% in all experiments, with absolute activity ranging up to 110 MBq (3 mCi). Future work will aim to optimise the beam energy to maximise activation yield and control effects of beam heating, and will explore alternative target materials and reactions. These results offer potential new production mechanisms for ^{18}F and complement the existing portfolio of tracer particles used in positron emission particle tracking experiments.

The authors wish to thank the iThemba LABS scientific programme advisory committee for enabling this work.

References

1. D.J. Parker, Rev. Sci. Instrum. 88, 051803 (2017)
2. T.W. Leadbeater, D.J. Parker, J. Gargiuli, Particuology, 10, 146-153 (2012)
3. R.D. Wildman, C.M. Hrenya, J.M. Huntley, Granular Matter, **14**, 215-220 (2012)
4. A.J. Morrison, I. Govender, A.N. Mainza, D.J. Parker, Chem. Eng. Sci., **152**, 186-198 (2016)
5. R.C. Sindall, D. Dapelo, T. Leadbeater, J. Bridgeman, Flow Meas. Inst., **54**, 250-264 (2017)
6. T. Leadbeater, J.P.K Seville, D.J. Parker, Canadian J. of Chem. Eng., 10.1002/cjce.24622 (2022)
7. A. Buffler, I Govender, J.J. Cilliers, et al., Int. J. Mod. Phys. Conf. Ser. 48, 1860113 (2018)
8. T. Leadbeater, A. Buffler, K. Cole, M. van Heerden, A. Camroodien, D. Steyn, Int. Top. Meeting on Nuc. App. Accl., 230-241, (2021)
9. T. Leadbeater, A. Buffler, M. van Heerden, A. Camroodien, D. Steyn, Nucl. Sci. & Eng. , **10.1080/00295639.2023.2171234** (2023)
10. A.J. Knoing, D. Rochman, J. Sublet, et al., Nuclear Data Sheets, **155**, (2019)
11. G.F. Steyn, C. Vermeulen, A.H. Botha, Nucl. Inst. Meth.:A, 727, 131-144 (2013)

Comparison of Bi_2O_3-targets produced by thermal evaporation and RF magnetron sputtering

Birgit Kindler[1*], *Bettina* Lommel[1], *Elif* Celik Ayik[1], *Annett* Huebner[1], *Jutta* Steiner[1], *and Vera* Yakusheva[1]

[1]GSI Helmholtzzentrum für Schwerionenforschung mbH, 64291 Darmstadt, Germany

Abstract. For the nuclear chemistry and for the nuclear physics of the heavy elements, bismuth is one of the key target materials, as it is the heaviest stable element. As discussed earlier, compound targets were developed to withstand higher intensive heavy ion beams. In the past, Bi_2O_3 was evaporated from the tantalum crucible and deposited on carbon backings. As the melting temperature of Bi_2O_3 is high (817° C), the process is at the limit of the cooling features of the evaporation set-up. Therefore, we decided to test RF magnetron sputtering as an alternative production method. We will present results of the different behavior of the targets produced via alternative processes.

1 Introduction

Bismuth is an important target material, especially in heavy and super-heavy element (SHE) experiments since it is the heaviest stable nuclei for the synthesis of odd compound nuclei with cold fusion reactions. In heavy ion beam experiments, the low melting temperature of the metal with 271 °C is a problem for high-intensity beams; therefore, it was necessary to investigate higher melting compounds as an alternative, as already done for lead compounds [1].

At first, the bismuth compounds were produced by thermal evaporation, which is a well-established method with a high yield. In addition, it is the method of choice for most of the targets for the super-heavy element program for decades. As bismuth is monoisotopic, this opened the possibility to test radio frequency (RF) magnetron sputtering as an alternative method, as will be described below.

2 Bismuth compounds

As alternative for metallic bismuth, we searched for simple binary compounds with higher melting temperatures. We could buy all compounds readily and deposited them by thermal evaporation on carbon backings on a rotating evaporation wheel, as described in [1]. As a standard, 350 - 450 µg/cm² of the target material was evaporated from a tantalum crucible on a carbon backing of 35 - 45 µg/cm², which is produced in our laboratory by resistance heating [2]. The following compounds were tested.

Bismuth sulfide (Bi_2S_3) has a melting temperature of ~ 685 °C. The compound already decomposed during thermal evaporation, so Bi_2S_3 was discarded right away. Bismuth fluoride (BiF_3) has a melting temperature of 727 °C. The fluoride could be evaporated successfully on carbon backing though the as prepared foils showed some stress. A stress relief by heating the foils from the backing side during evaporation, which often improves the quality and stability of foils, in this case only led to decomposition of the material. Bismuth oxide (Bi_2O_3) has a melting temperature of ~ 817°C. Since the melt is very corrosive, a platinum crucible had to be used for evaporation in this case. The bismuth oxide on carbon backing could be evaporated successfully with heating of the backings as well as without heating. Targets from both production routes were prepared

2.1 Final choice of Bi_2O_3 as Targets

In 2003, test experiments at the heavy-ion separator SHIP (Separator for Heavy Ion Particles) were performed with the described compound targets [3]. All compound targets showed cross-sections comparable to that of the metal, but the oxide targets exhibited a much higher durability. Oxide targets formed on heated backings were the most durable. In addition, the melting temperature of the oxide is higher than that of the fluoride. Therefore, the oxide was the alternative of choice. Such targets were applied since as standard bismuth in several beam times. Bismuth oxide is now the target of choice. Some of the stored targets, however, showed peculiar degradation during storage for an unknown reason. Therefore, alternative methods were sought for achieving better properties of Bi_2O_3-targets.

3 Properties of bismuth oxide

Bi_2O_3 is a polymorphic system with four known different crystallographic phase were the α- and the δ-phase are stable, and the β- and γ-phase are metastable phases only formed upon cooling [4]. Since the high-

* Corresponding author: b.kindler@gsi.de

temperature δ-phase has a very high ion conductivity, it is a good candidate as a solid electrolyte for fuel cells or in sensors [5]. Therefore, over the last decades there were intensive investigations on the phase transitions in this binary system with various methods; see e.g. the paper of Schröder et al. [6] and citations therein. Depending on the cooling rate, the atmosphere, the pressure and other parameters, different crystallographic structures occur and are probably frozen. Klinkova et al. [7] also reported oxygen deficiency of a varying extent for the different phases.

With thermal evaporation, there is no possibility to control the cooling and condensation of Bi_2O_3 target material on the backing. Quenching of metastable phases could be possible explanations for the stress within the deposited layers responsible for instability for some targets.

To avoid degradation of the metastable Bi_2O_3 and the restructuring of carbon backings during the production or storage, an alternative production method was sought in which the metastable phases could be avoided. This could lead to better-defined phase homogeneity in the deposited Bi_2O_3 layers. Unfortunately, we could not detect different phases with the methods available.

4 RF Magnetron sputtering

Since magnetron sputtering is a ballistic coating process, the liquid phase and the undesired β- and γ-phase can be avoided. Bismuth is monoisotopic, thus easy and relatively inexpensive to buy as a sputter target. The higher losses compared to thermal evaporation are not a crucial point, as it would be for isotopically enriched target material. Bismuth oxide is a poor electric conductor, so radiofrequency sputtering (RF) magnetron sputtering allows for a stable plasma process.

With RF magnetron sputtering, Bi_2O_3 targets were produced on carbon backings without major problems, though the foils in some batches were heavily stressed and broke already during production. While it is not yet understood, why this sometimes happens, the surviving foils are stable during storage and show no degradation with time. The method and set up used for magnetron sputtering is described in detail in [8, 9].

5 Investigations

To find differences between targets produced with the different techniques, selected targets were analysed. We investigated targets produced with both methods that were not irradiated and that showed no obvious degradation; targets produced with both methods that were irradiated with ^{48}Ca-ions during a SHIP beam time; targets that were not irradiated but showed peculiarities nevertheless.

For this paper, as a first step, only tools and methods were applied that were available in our laboratory to investigate the selected samples. From each target investigated, a photograph of the whole target on the frame was taken to see any phenomenological differences or peculiarities. In addition, a quantitative analysis with an Energy Dispersive X-ray detector (EDX) was performed, evaluating the ratio of bismuth to oxygen in units of atomic percentage for each sample. We report this ratio for each sample picture in a green box. This value is the calculated mean value of four different point measurements in the region of the target, which is indicated by the red arrow. For non-irradiated targets, only the centre region was investigated. For targets irradiated with an ion beam, the centre region was measured representing the irradiated area and the peripheral area where the beam intensity was minimum for comparison. We only had targets at our disposal that were non-irradiated leftover reserves from beam time campaigns or irradiated targets that survived the ion beam bombardment. Therefore, the number of available specimen was largely restricted. The measured ratios show a rather wide scatter so that only trends are observed and no systematic reliable information.

As a reference, a tablet of the starting material of Bi_2O_3 was analysed by EDX in the same way as the foils, giving a ratio of Bi/O of 0.9. All targets investigated had a thickness between 400 and 500 μg/cm² referring to the bismuth content and carbon backings with a thickness between 40 and 60 μg/cm² for those prepared with thermal evaporation and between 40 and 50 μg/cm² for those prepared by magnetron sputtering.

5.1 Targets prepared by thermal evaporation

All targets shown in figures 1 – 5 were prepared by thermal evaporation. The targets in figure 1 and 2 were not irradiated in a beam time.

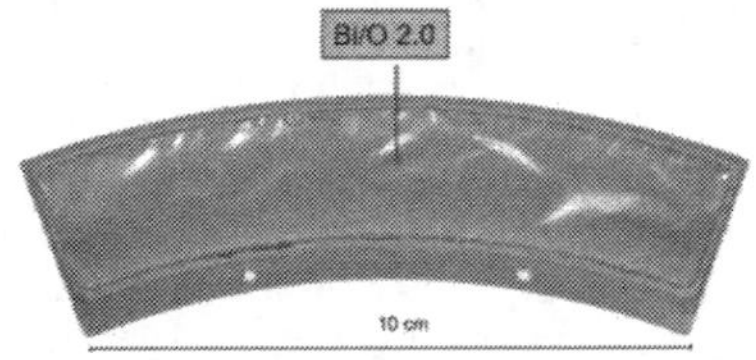

Fig. 1. #7-Bi_2O_3 on C-backing by thermal evaporation on SHIP-frame, not irradiated. Green box indicates ratio of atomic % evaluated by EDX.

In the target shown in figure 2, the carbon backing partly restructured during the evaporation from a smooth layer to a corrugated structure. This effect is a well-known effect for carbon layers. It happens spontaneously upon heating and cannot be influenced by the operator. However, the restructuring seems not to influence the performance during beam time experiments. The Bi/O-ratios here vary between 2.0 and 2.7.

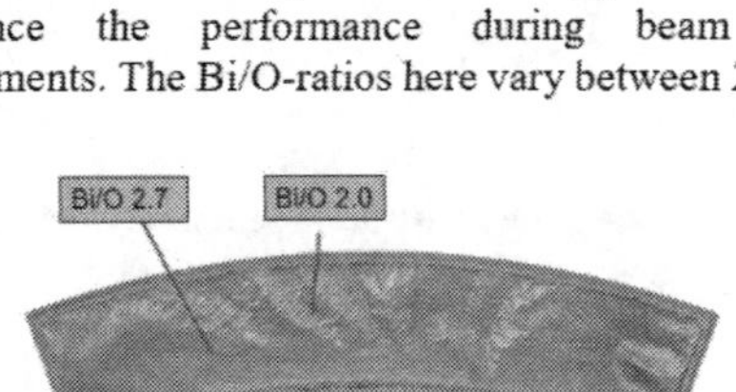

Fig. 2. #5-Bi_2O_3 on C-backing by thermal evaporation on SHIP-frame. Carbon backing partially restructured during evaporation, not irradiated. Green box indicates ratios of atomic % evaluated by EDX.

The target shown in figure 3 was also not irradiated but was stored for 3 years in air. The whole batch of targets that were prepared in that one run disintegrated with time. Nevertheless, this did not happen for other batches that were stored even for longer times. The Bi/O-ratio of 0.9 in this case is noticeable lower than for the intact targets.

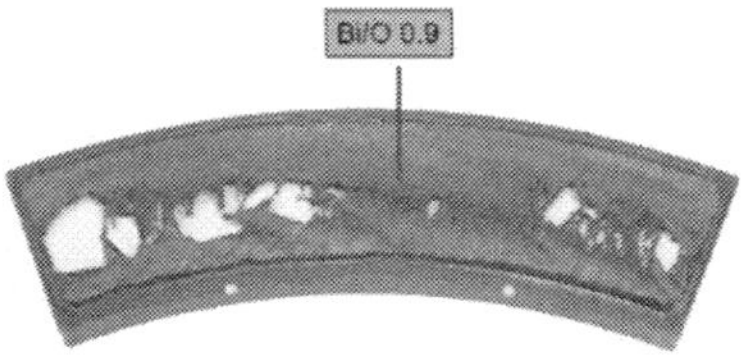

Fig. 3. #79-Bi_2O_3 on C-backing prepared by thermal evaporation on SHIP-frame in 2018 and stored in air for 3 years. Green box indicates ratio of atomic % evaluated by EDX.

The targets shown in figures 4 and 5 were irradiated in a SHIP beam time with ^{48}Ca-ions. The intensity distribution of the beam profile during the irradiation is clearly visible. For the target in figure 4, the profile was not well distributed resulting in distinct colour changes and corrugations in a small strip along the centre of the target were the intensity and consequently the temperature was highest. The Bi/O-ratio at the less-irradiated periphery in both cases are 0.9, which is the ratio in the reference material.

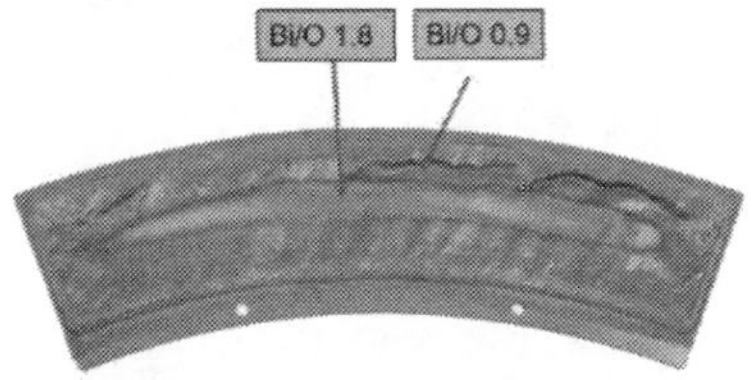

Fig. 4. #15-Bi_2O_3 on C-backing prepared by thermal evaporation on SHIP-frame and irradiated with a ^{48}Ca-beam (~ $2*10^{17}$ particles). Green box indicates ratios of atomic % evaluated by EDX.

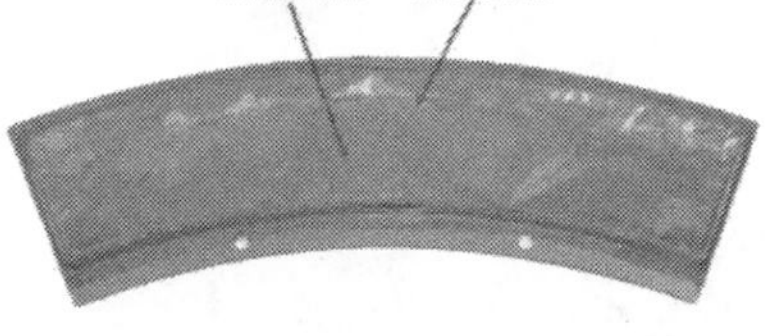

Fig. 5. #29-Bi_2O_3 on C-backing prepared by thermal evaporation on SHIP-frame and irradiated with a ^{48}Ca-beam (~ $4*10^{17}$ particles). Green box indicates ratios of atomic % evaluated by EDX.

5.2 Targets prepared by magnetron sputtering

The targets shown in figures 6 – 8 were prepared by RF magnetron sputtering; those in figure 6 and 7 were not irradiated.

Sometimes, all targets of one batch show strong tensions so that the layers tear or even burst away during production, as can be seen in figure 7. We cannot influence or predict this effect. The Bi/O-ratios seems to be overall lower comparable to those observed in the targets prepared with thermal evaporation but this is not significant.

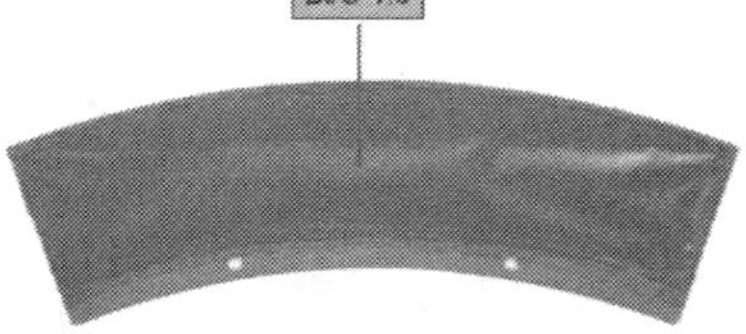

Fig. 6. #15SP-Bi_2O_3 on C-backing prepared by magnetron sputtering. Green box indicates ratio of atomic % evaluated by EDX.

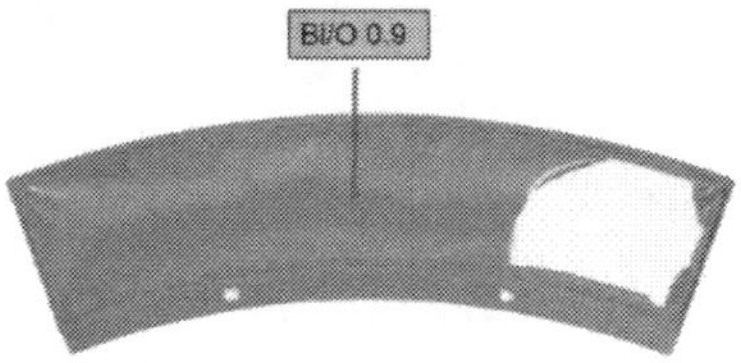

Fig. 7. #21-Bi_2O_3 on C-backing prepared by magnetron sputtering - foils show high tensions. Green box indicates ratio of atomic % evaluated by EDX.

The target shown in figure 8 was irradiated with a ^{48}Ca-beam. In addition, here we can observe a clear difference in the Bi/O-ratio between the periphery of the target and the centre, where the beam intensity was highest.

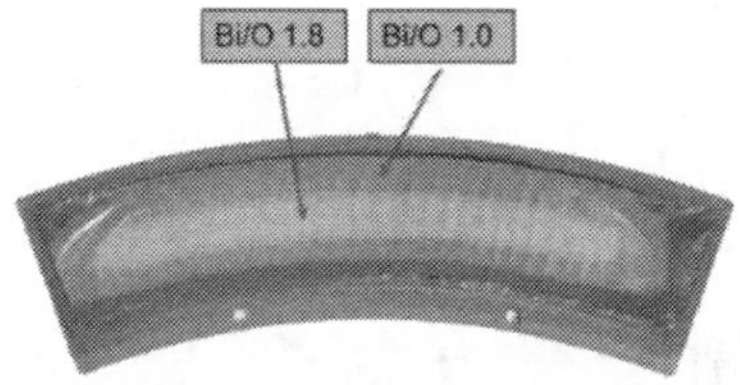

Fig. 8. #50-Bi_2O_3 on C-backing prepared by magnetron sputtering and irradiated with ^{48}Ca ($2*10^{18}$ particles). Green box indicates ratios of atomic % evaluated by EDX.

5.3 Targets on titanium backing

In figure 9, targets are shown that were produced very recently for a special experiment where a titanium backing was required. To find the most appropriate production process for this experiment, we prepared targets of the same thickness with the different processes for a direct comparison. The titanium backing had a thickness of about 2.1 µm in all cases. The two targets on the left in figure 9 were prepared by thermal evaporation; the target on the far left was prepared without heating of the backing during evaporation (figure 9a) and for the target in the middle (figure 9b), the backing was heated during evaporation. The target in figure 9 on the right was prepared with RF magnetron sputtering (figure 9c).

Here, we see significant differences already in the appearance of the three foils. Rather striking are the Bi/O ratios here. Whereas the Bi/O-ratio of the sputtered target of 1.4 lies well within the range observed for all the targets prepared on carbon backing, the ratio of 4.5 for both targets prepared by thermal evaporation are significantly higher.

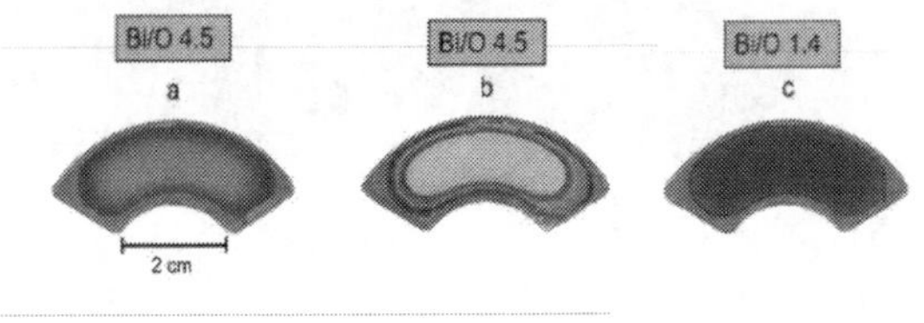

Fig. 9. Bi_2O_3 on titanium backing prepared by
a: Thermal evaporation without heating of the backing,
b: Thermal evaporation with heating of the backing
c: Magnetron Sputtering
Green box indicates ratios of atomic % evaluated by EDX.

6 Discussion and interpretation

The Bi/O-ratios for all targets with carbon backing vary in the range from 0.9 to 2.7. We could neither observe any systematic difference between the foils prepared by thermal evaporation and magnetron sputtering nor between original and irradiated foils. Compared to the Bi/O-ratio of the starting material Bi_2O_3 of 0.92, most layers are largely oxygen deficient. The performance in beam time experiments was comparable though there is some evidence that sputtered targets might show a longer durability.

Though there is evidence from the evaluated Bi/O ratios that bismuth is lost in the beam spot region during irradiation, there was no drop in reaction cross sections reported from the experiment and the targets did not lose weight during irradiation.

For layers deposited on Ti-backing, we see significant differences in the Bi/O-ratio between the two methods.

Already the appearance of the foils is quite different. The measured Bi/O ratios are interesting. For the targets produced with thermal evaporation, the Bi/O ratio is significantly higher than those observed for targets on C-backing: 4.5 compared to max. 2.7. This is not observed for targets prepared by magnetron sputtering.

Because of the complicated phase transformations of Bi_2O_3 upon cooling the structure of evaporated layers is not controllable with our techniques. Presumably, this should happen less in sputtered foils where no liquid phase is involved, but this is not obvious for the investigated samples. However, the large amount of the observed oxygen loss cannot be explained solely with a phase mixing. Such a significant loss of oxygen in preparation could have a stoichiometric background or could be due to different sticking coefficients of oxygen and bismuth on the substrate. Different sticking behaviour probably is the major factor for the differences observed between layers deposited on carbon and on titanium backings. However, as long as the bismuth atoms remain in an oxygen binding that raises the melting temperature sufficiently, the durability and performance during irradiation experiments seems not to be affected much by the actual structure or oxygen content. In the case of titanium backing, the titanium might act as a getter material for the oxygen thus reducing the oxide especially at elevated temperatures. This might be avoided with magnetron sputtering.

7 Outlook

For better statistics, more beam time experiments are planned with comparable beam conditions to identify possible differences in the durability of foils produced with the different methods. Preferably, each target set should be irradiated until a defined degradation state is reached. This has to be negotiated well with the scientists performing the experiments since beam time is precious and rare. Furthermore, the degradation on air with time for foils with alternative methods in order to understand the mechanism behind the effect will be observed. The targets produced on titanium backing have to be tested in a beam time to see if their durability differs significantly from those deposited on carbon backings. Also very important will be to investigate the layer composition of targets produced with different methods with analytic methods more sensitive to the chemical bonding and quantitative composition of the constituents, as e.g. infrared or Raman spectroscopy.

References

1. B. Lommel, D. Gembalies-Datz, W. Hartmann, S. Hofmann, B. Kindler, J. Klemm, J. Kojouharova, J. Steiner, Nucl. Instr. and Meth. A 480, 16-21 (2002).
2. B. Lommel, W. Hartmann, B. Kindler, J. Klemm, J. Steiner, Nucl. Instr. and Meth. A **480**, 199-203 (**2002**).
3. B. Kindler, D. Ackermann, W. Hartmann, F. P. Hessberger, S. Hofmann, B. Lommel, R. Mann, J. Steiner; Nucl. Instr. and Meth. A 561, 107-111 (2006)
4. H. A. Harwig, A. G. Gerards, Termochimica Acta 28 (190) 121-131.
5. A. M. Azad, S. Larose, S. A. Akbar, J. of Mat. Sci. 29 (1994) 4135–5151
6. F. Schröder , N. Bagdassarov, F. Ritter, L. Bayarjargal, Phase Trans. 83:5, 311-325 (2010)
7. L. A. Klinkova, V. I. Nikolaichik, N. V. Barkovskii, V. K. Fedotov, Russ. J. of Inorgan. Chem., 52, No. 12, 1822–1829 (2007)
8. B. Kindler, W. Hartmann, J. Klemm, B. Lommel, J. Steiner, Nucl. Instr. and Meth. A 521, 222-226 (2004)
9. B. Kindler, W. Hartmann, A. Hübner, B. Lommel, J. Steiner, V. Yakusheva; J. of Radioanalyt. and Nucl. Chem. 305, 703-706 (2015)

The preparation of isotopic boron targets – searching for a more consistent approach

John Greene[1*] and *Matthew* Gott[1,2]

[1]Physics Division, Argonne National Laboratory, 9700 South Cass Avenue, Argonne, IL 60439 USA
[2]Enrichment Science and Engineering Division, Oak Ridge National Laboratory, 1 Bethel Valley Road, Oak Ridge, TN 37830 USA

Abstract. The reliable availability of isotopic boron targets has gained in importance concerning light ion reactions withing the realm of nuclear astrophysics research and remain somewhat elusive. An exhaustive approach was undertaken via electron beam evaporation to produce self-supporting and backed targets spanning quite a few isotopic samples across several suppliers, resulting only in limited success. Details of the extensive sample preparation procedures undertaken, and the experimental deposition techniques explored will be presented.

1 Introduction and Motivation

For ongoing experiments in low-energy nuclear physics at the Argonne National Laboratory (ANL) Argonne Tandem Linear Accelerator System (ATLAS) Facility and especially studies involving nuclear astrophysics, isotopic boron targets are becoming more in demand. While current preparation techniques involve Physical Vapor Deposition (PVD) employing electron beam heating [1], reliable and consistent production of targets has been difficult to achieve. Here, the approach has been to research and document a standard procedure capable of consistently obtaining the targets needed for meeting the experimental demands.

There are many references on research pertaining to preparing boron films via the process of electron beam heating, beginning with Eschbach in 1972 [2] as well as others [3,4,5,6]. The results presented here extend the earlier work of Thomas [1,7]. Various alternate methods also appear in the literature, including electrodeposition by Verdingh [8] and Pauwels [9], focused ion beam sputtering by Baumann and Wirth [10], and Maier [11] and Muggleton [12]. Xu and Wang [13] employed the method of centrifugal precipitation, and Lozowski and Hudson [14] prepared thick targets via powder pressing. A new technique, High Energy Vibrational Powder Plating (HIVIPP), developed by Sugai [15] was used to produce boron films and later by Lipski, et al [16]. It should be noted that early on, boron was being considered as an alternative to carbon for use as stripper foils by many researchers, including Ramsay [17], and with Zeisler and Jaggi [18], incorporating multi-layers [19] and via laser ablation deposition [20].

Finally, a lot of effort has gone into the characterization and assay of boron film layers, for instance, as neutron dosimeters [21], at the Institute for Reference Materials and Standards as reference standards for nuclear measurements [9,22,23,24] and for neutron physics experiments [24,26] at the National Institute of Standards and Technology.

2 Isotopic Starting Material

Boron is found in nature as compounds, primarily borax, the element being discovered in 1808 by Sir Humphry Davy. High-purity crystalline boron is obtained from the vapor phase reduction of boron trichloride with hydrogen on electrically heated filaments or via exchange distillation reactions [27] using a boron trifluoride compound (U.S. Patent 3,050,367). Its melting point is 2300 °C, but it sublimes at 2550 °C. Boron exists as two isotopes, ^{10}B 19.78% abundance and ^{11}B with 80.22 % abundance. The isotopes are available as crystalline form in high purity from Oak Ridge National Laboratory (Oak Ridge, TN, USA) and also from Eagle-Picher (EP Boron Quapaw, Oklahoma).

There exist at our disposal several legacy batch samples of various enrichments for both isotopes (age unknown) for use in preparing accelerator targets for ATLAS. Detailed assay and chemical analysis obtained for one of the ORNL enriched ^{10}B samples were given to be 98.4% total boron with 1.20% carbon and 0.28% oxygen along with other listed ppm impurities, the largest being H_2O, Fe, Si and S. A similar analysis provided for one of the ^{11}B samples showed 98.2% total boron with 1.23% carbon and 0.10% oxygen and smaller ppm impurities of H_2O and Si. A lone sample of ^{nat}B powder (purity 99.995%) from Lieco Industries (Lyndhurst, NJ, USA) was included for comparison to the isotopes. The high-purity of this material would signify only low-levels of impurities to be present.

Recent work by De Luna *et.al.* characterized similar ^{10}B powder (97%) obtained from Quapaw, OK using a variety of techniques including FTIR, SEM, XRD, and

* Corresponding author: greene@anl.gov

TOF-SIMS and compared it to natural boron [28]. The results for the enriched ^{10}B powder revealed 2% mass content of boric acid as well as other minor constituents. Anecdotal evidence as well as experience in the preparation of accelerator targets has shown that, at times, deposition of natural material behaves in a fashion completely different from that of the enriched isotope. This is particularly true for the case of boron both in the production yield (i.e., non-adherence to the substrate as well as marginal success in floating foils) and ultimate thicknesses achieved. These difficulties in obtaining self-supporting foils from the isotope may possibly arise from the chemistry pathways employed during the enrichment process and thus differ from the naturally occurring starting material. A further example is the fact that crystalline boron isotope upon heating exhibits a tendency to eject from the electron beam hearth – due perhaps to volatile compound decomposition contained in the solid isotope [1]. It has been reported by Nandi *et al.* [29] that amorphous, and to a lesser extent, crystalline boron powder tends to absorb water vapor over time. This was empirically determined from various aged samples and proposed vacuum flushing with an inert gas (Ar) to remove this trapped H_2O content. De Luca *et.al.* [28] proposed drying the boron material in a vacuum oven and purged with N_2 at 130 °C. IR spectra obtained for amorphous ^{nat}B and crystalline ^{10}B before and after heating were consistent with results from Nandi *et al.* with the crystalline material showing less boric acid than the amorphous powder.

3 Experimental Procedures

As per the earlier work of Thomas at ANL [1], the powdered starting material is first pre-heated to drive out any volatile material. This heating was carried out withing the Intlvac Evaporator System [30] using a Ta crucible heated resistively from a crucible oven source. The powder was first pressed into a 0.95 cm diameter pellet using a hand press. The source is brought to approximately 1300 °C, measured using an optical pyrometer and heated until outgassing ceases (as registered by the nominal system pressure; base pressure ~ 10^{-7} Torr), and stopped when deposition commences.

Table 1. Sample pre-heating under vacuum

SAMPLE	WEIGHT BEFORE (mg)	WEIGHT AFTER (mg)	%WEIGHT LOSS
^{nat}B	109	90	17
^{nat}B	202	180	11
	88	81	8
^{nat}B	203	177	13
	202	176	13
	199	172	14
^{10}B SS 1(a) 140 1210-1	310	291	6
^{10}B #4700117 24-3177	428	385	10

The pellet was left to cool. It had been weighed prior to the heating and then re-weighed after removal afterwards. Table 1 provides results of several such heat-treated samples. Most of the powdered samples lost weight on the order of 15%.

3.1 Sample Preparation

Another aspect of standardization of the boron starting material is particle size. Using a calibrated set of sieves, samples were prepared with a particle size of 60 mesh (upper limit). This technique works fine for powdered samples, in fact, the ^{nat}B powder described above was found to be finer (80 mesh). However, most of the isotopically enriched materials (both ^{10}B and ^{11}B) were crystalline form. To prepare uniform powder starting samples from this crystalline form, the large crystalline chunks were first crushed using a percussion mortar and pestle. This material was further reduced in particle size using a SPEX Sample Prep ball mill and Ta ball media. For all samples, the resulting powder sample is then passed through a 60-mesh sieve, pressed into a pellet and pre-heated to remove volatile impurities.

To avoid loss of sample due to evaporation during vacuum heating, an alternate technique of pre-heating the isotopic powder samples was carried out in the target laboratory tube furnace under flowing Ar gas (1-2 cc/min). As before, the isotope powder was compacted into a pellet and heated to 1000 °C for 1 hour in a Ta crucible. The samples were weighed before and after. The results are given below for $^{10,11}B$ in Table 2. Due to the abundance of isotopic powder, an additional two ^{10}B samples #4700117 (ORNL) were also heated by this method and showed significantly higher weight loss after pre-heating than when heated under vacuum.

Table 2. Sample pre-heating in tube furnace at 1000 °C for 1 hour

SAMPLE	WEIGHT BEFORE (mg)	WEIGHT AFTER (mg)	%WEIGHT LOSS
Eagle-Picher 1279 5010026 (^{10}B)	406	380	6
#4700117 24-3177 (^{10}B)	407	332	18
	403	324	20
SS 1a-140C 29-1977 (^{10}B)	408	395	3
SS 2(ce) Req. 1506-204 (^{11}B)	398	370	7
	403	371	8
ENGLAND #22 20th Century Electronics (^{11}B)	407	377	7
	407	382	6
Eagle-Picher B-11-85-1 (^{11}B)	409	405	1
	397	394	<1

3.2 Substrate Preparation

The substrates chosen for this work are standard laboratory microscope slides, unfrosted, with dimensions 2.5 cm x 7.5 cm, obtained from Fisher Scientific Inc. (Waltham, MA, USA). Approximately 50-60 μg/cm^2 of NaCl are evaporated onto the slides similarly from a Ta crucible in a resistive heating crucible oven source, also contained within the Angstrom Deposition Tool [31] (see Fig. 1). From experience, sodium chloride was chosen over boron oxide as suggested by Ramsay [32]. For this substrate layer, four clean glass slides are placed above the evaporation source at 11 cm. The quartz crystal deposition monitor was positioned at 17 cm away, centered directly above the Ta crucible source.

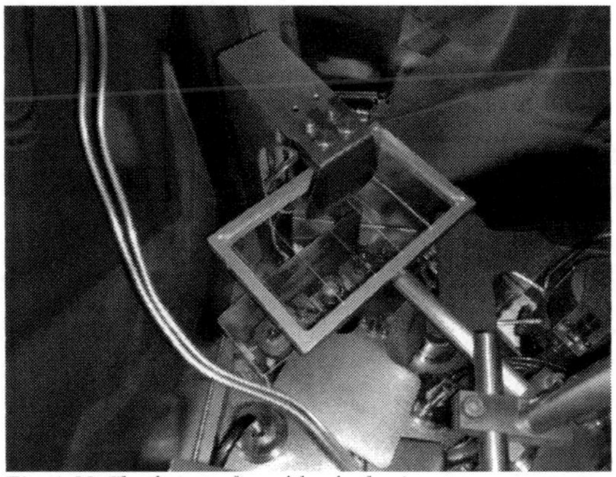

Fig. 1. NaCl substrate deposition in the Angstrom evaporator

3.3 Electron Beam Source

The boron evaporations (natural and isotopic) were carried out via physical vapor deposition (PVD) within the Angstrom Deposition Tool using the modified Telemark electron beam source (see, for example [33]). The natural boron was employed to carry out an initial test evaporation using deposition parameters previously determined for boron over many years at the ANL target laboratory. In a quick fashion, several self-support targets of thickness 100 μg/cm2 were easily obtained from this single evaporation, thus reaffirming the proper deposition data. No further deposition attempts with ^{nat}B were necessary.

Four previously prepared NaCl coated glass substrates were placed at 12 cm distance above the source and are direct heated from a halogen quartz lamp placed above. The substrates are held at a temperature of 176 °C as indicated from the previous work by Thomas and recommended by Gursky [34]. The indicated temperature is monitored via a thermocouple placed on the back of the substrate slides and the evaporator system maintains temperature by varying the output of the lamp. The boron sample is placed in one of the four available Cu hearths of the electron beam source, having first been recompacted into a fresh pellet via hydraulically pressing under vacuum to reduce outgassing during heating.

For these depositions an arbitrary endpoint of 100 μg/cm^2 target thickness was chosen to begin as these would provide usable targets. Thicker depositions could be performed later once the standardized parameters for this work had become established. Once under vacuum and the substrate heated, the electron beam power is slowly increased. With a beam voltage of 8-9 kV a deposition rate of 0.2 kÅ/s was slowly achieved and maintained throughout as measured by the quartz crystal monitor placed to the side at a fixed 22 cm distance from the source. Deposition was stopped when the 100 μg/cm^2 endpoint was reached and the system allowed to cool, usually overnight, before the slides were extracted.

3.4 Carbon Backing Foils

Carbon backings of 20, 30, and 40 μg/cm^2 were prepared on the various target frames anticipated as being needed. They were made using the standard floating technique from carbon slides obtained from Arizona Carbon Foil Company (www.acf-metals.com) as they have proven over time to be a consistently reliable source of high-quality carbon foils for use in our laboratory. For these evaporations, the deposition parameters were kept like those used for the self-support boron films.

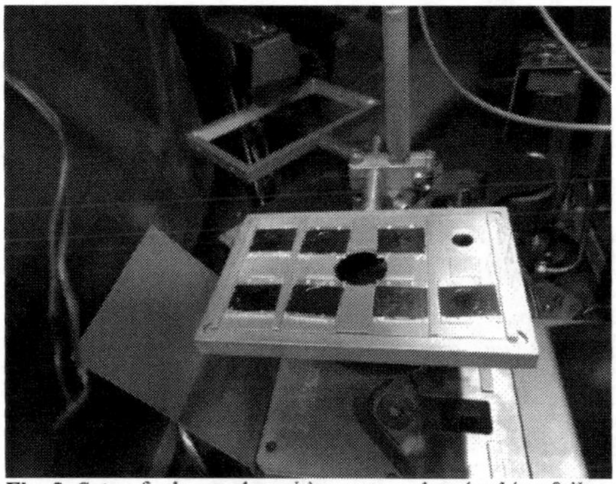

Fig. 2. Setup for boron deposition onto carbon backing foils.

4 Results and Current Progress

The thicknesses and number of successfully produced isotopic targets are presented below in Table 3. Compared to the relative ease of producing self-supporting foils from the natural boron, these enriched targets were only obtained after several attempts using many substrates. Commonly, the failure was due to stresses in the foils during evaporation; the foils would curl and flake apart, likely due to incomplete bonding between the layers. Due to the limited success with producing self-supporting ^{10,11}B targets, further deposition attempts were then continued using carbon backing foils (see Fig. 2; Table 4).

Certainly, disappointing was the fact that even after pre-heating and particle size uniformity, the isotopic target production difficulties were still present compared to natural boron. The limited success outlined here means there is still progress to be sought out. The possibility of heating the powder to higher temperature for a longer duration and analytically monitoring for boric acid, may provide an avenue for improved deposition behaviour.

https://doi.org/10.1051/epjconf/202328510002

Table 3. Isotopic boron self-supporting targets produced.

SAMPLE	THICKNESS (μg/cm²)	NO. OF TARGETS
SS 1(a) 140 1210-1 (^{10}B)	110	1
#4700117 No. 23-3177 (^{10}B)	56	4
Eagle-Picher #5010026 Lot No. 1279 (^{10}B)	101	12
SS 1a-140c No. 29-1977 (^{10}B)	83	3
ENGLAND 20th Century Electronics (^{11}B)	101	2
Eagle-Picher B-11-85-1 (^{11}B)	61	2

Table 4. Isotopic boron targets produced on carbon backings.

SAMPLE	THICKNESS (μg/cm²)	CARBON BACKING (μg/cm²)	NO. OF TARGETS
#4700117 24-3177 (^{10}B)	141 137 113 107 84	30 40 40 40 30	2 1 1 1 2
#4700704 No.26- 2578-B (^{11}B)	121	40	8
Eagle- Picher B-11-85-1 (^{11}B)	81	40	5

5 Conclusion

In conclusion, although multiple preparation techniques were explored for many samples of both isotopes, no direct correlation was ascertained as to successful target preparation. Pre-heating, particle size and powder grinding the boron samples provided a good approach, but the qualitative difficulties experienced with isotopic target preparation remain as compared to natural boron. It still appears that the chemical processing differences existing for the isotopic samples make producing self-supporting films difficult to achieve. It was desired that among the many isotopic samples explored, procedures could be developed to alter their properties for producing thin target foils. There will likely be a steady demand for these targets once reliable preparation techniques can be established. Further investigations are therefore warranted.

This material is based upon work supported by the U.S. Department of Energy, Office of Science, Office of Nuclear Physics, under Contract No. DE-AC02-06CH11357. This research used resources of ANL's ATLAS facility, which is a DOE Office of Science User Facility.

References

1. G.E. Thomas, NIMA 282 (1989) 124
2. H. Eschbach, NIM 102 (1072) 469
3. L. Csihas, NIM 167 (1979) 171
4. W.D. Riel, NIM 167 (1979) 179
5. Xu Guo Ji and Wei Yong-Qin, NIMA 397 (1997) 62
6. Y. Sasamoto, T. Kawabata, M. Hamagaki, I. Sugai and Y. Takeda, NIMA 590 (2208) 66
7. G.E. Thomas, NIM 200 (1982) 27
8. V. Verdingh, Report AERE-R 5097 (1965) 58
9. J. Pauwels, NIMB 56/57 (1991) 938
10. H. Baumann and H.L. Wirth, NIM 167 (1979) 71
11. H.J. Maier, NIMA 303 (1991) 172
12. A.H.F. Muggleton, NIMA 334 (1993) 3
13. G. Xu and C. Wang, NIMA 236 (1985) 555
14. W.R. Lozowski and J.D. Hudson, NIMB 79 (1993) 845
15. I. Sugai, NIMA 397 (1997) 81
16. A.R. Lipski, G. Rainovski, N. Pietralla and A. Dewald, NIMA 590 (2008) 69
17. D. Ramsay, Report LA-6850-C (1977) 74
18. S.K. Zeisler and V. Jaggi, NIMA 613 (2010) 434
19. V. Jaggi, S. Olson, B. Kumar and S.K. Zeisler, J. Radioanal. Nucl. Chem. 299 (2014) 1019
20. S.K. Zeisler, M. Brigham, I. Kaur and V. Jaggi, AIP Conference Proceedings 1962, 030015 (2018); https://doi.org/10.1063/1.5035532
21. W.D. Box, NIM 102 (1972) 43722
22. J. Van Audenhove, V. Verdingh, H. Eschbach and P. Debievre, Proceedings (1974) 119
23. H. Tagziria, et.al., NIMA 303 (1991) 123
24. J. Pauwels, et.al., NIMA 362 (1995) 104
25. J. Pauwels, et.al., NIMA 303 (1991) 133
26. D.M. Gilliam, G.P. Lamaze, M.S. Dewey and G.L. Greene, NIMA 334 (1993) 149
27. Handbook of Chem. And Phys. 56th Ed., R.C. Weast, 1974,1975 by C.R.C. Press, Inc.
28. De Luna *et.al.,* NIMA 1000 (2021) 165232
29. Nandi *et.al.,* Cent. Eur. J. Energ. Mater. 9 (4) (2012) 387-398
30. J. P. Greene, J. Neubauer and D. Deligiannis, NIMA 561 (2006) 58
31. John P Greene and Mina Labib, AIP Conference Proceedings 1962, 030018 (2018); https://doi.org/10.1063/1.5035535
32. D. Ramsay, Proceedings (1974) 151
33. G.E.Thomas, J.P.Greene, P.Maier-Komor and R.H.Leonard, NIMA 303 (1991) 162
34. J.C. Gursky, Report LA-6850-C (1977) 198

AUTHOR INDEX

AUTHOR INDEX

EDP Sciences – Web of Conferences
17, Avenue du Hoggar
Parc d'Activité de Courtabouf
BP 112
F-91944 Les Ulis Cedex A
France

ISBN 978-1-7138-8042-4

CURRAN ASSOCIATES INC.
proceedings.com

International Conference on Advanced Functional Materials and Composites (ICAFMC 2018)

MATEC Web of Conferences Volume 242 (2018)

Barcelona, Spain
5 - 6 September 2018

edp sciences